SDN

Defining a Strategic, Business-focused Architecture

James J. Connolly

1st Edition, 2015

ISBN-13: 978-1508542834

Edited by Sonja Ruile; sonjaruile@gmail.com

Cover Design by Ashley Gordon, theashleygordon.666@gmail.com

About the Author

James J. Connolly has worked in various sectors across the telecoms industry over the last 20 years, with a significant amount of time dedicated to developing services and infrastructure solutions within the cable industry. His roles have included Director of Network Design/Architecture and Director of Technology Strategy at the largest international cable company.

James is a passionate advocate of SDN. With a crystal-clear focus on simplification and the customer, James has shared the concepts he has worked on over the past 10 years with vendors, at conferences and with network operators, in order to fuel innovation and change. In conjunction with various parties, many of these ideas have been further developed, and are now being adopted for use within the network operator's wider architecture.

Acknowledgements

Writing and sharing my thoughts on how I feel the evolving technologies will positively affect and shape the future of the telecoms industry has been a great opportunity and experience. However, bringing this book together has only been possible because of countless conversations, discussions and arguments over very many years. For this, I would like to acknowledge the following people who, through their input and sharing of their thoughts, corrected my misunderstandings or were willing to be influenced by my thinking. In some cases, the best input was when you told me I was wrong, and because of this I found other ways to better express my thoughts. This has been the key reason as to why I chose to write this book from both a business and technology point of view.

Acknowledgement of how you all shaped and redirected my thinking goes out to Wim Henderickx, Phil, Erwin Karlas, Berislav Todorovic, Steven van Steen, Eric Lennon, Mourad Veeneman, Rene Huizinga, Dirk Van Den Borne, Mark Whiteway, Steve Iatrou, Jeff Finkelstein, Chris Donley, Patricia Loppé, Gert Grammel, Chris Hughes, Bruno Rijsman, Brian, Cor Zwart and Steve Opferman - with massive appreciation! In addition, my thanks go out to Ashley for the cover design and Sonja for her patience and ability to find sentence structure where sometimes none existed.

For those I forgot: I acknowledge your contributions and apologise for not including you. In addition, I would like to say thanks to all across the SDN and NFV community for the excellent and exploratory work they have done, are doing and will continue to do as they continue to reshape how the industry not just operates, but also perceives itself.

Lastly and firstly, my thanks and appreciation go out to Natasja. Thank you for your support, for creating the time for me, and for listening to my thoughts throughout the process of writing this book. It simply wouldn't have happened without you.

Thanks again everyone.

"You cannot solve a problem from the same consciousness that created it. You must learn to see the world anew."

"Whoever undertakes to set himself up as a judge of truth and knowledge is shipwrecked by the laughter of the gods."

Albert Einstein

Table of Contents

Preface

Many early-adopter network operators are already benefiting from the advantages brought by SDN. This book discusses the additional values SDN can deliver to and how the goals of SDN create change, through the whole network operator organisation.

Expanding upon two papers that were written by the author and that were shared and discussed with industry experts within the last five years, the goal of this book is to describe how a business-focused strategic architecture can be created through combining the technology developments of Cloud, SDN, NFV, Big Data and Next Generation OSS. It reviews many of the needs of the network operators and explains an end-to-end architecture in which SDN can become the linking technology to enable a strategic business-focused architectural solution for network operators.

It is important to note that the term "network operator" is used in this book to describe any company that controls and operates a telecommunications network, with the aim of serving their customer base (e.g. Web Platform, PTT, Cable, Mobile, Enterprise, OTT, etc.) This generalised term is used because all these business models experience the same limitations with the existing technology approach.

As this is a strategic architecture, not all the components and protocols are readily available in the market place or are fully ratified. However, a lot of what is discussed in this book is either available or is currently in an early stage of development by the appropriate parties. To further enhance the architecture, suggestions are made for the creation of new management and control applications, which, like the architecture itself, should be created with the focus of delivering to the customer. By publishing this book before the technologies are finalised in their development, the author hopes to share his thoughts and experiences and to enable others to consider this strategy, as they drive the future developments to meet the needs of the network operator.

This will allow the architects, designers and engineers to drive their company's strategy and to ensure the delivery of an architectural solution that will address the individual and specific business goals of their company.

How to read 'SDN Strategic Architecture'

This book gives an overview of the many new capabilities that become possible, as this technology evolves to create a technology solutions architecture that delivers against the individual business goals of a network operator.

The initial chapters in this book introduce and explain many of the concepts to the reader. These sections provide key information to the reader about these new technologies, the business reasons as to why change is needed, and the different working approaches that can be considered through moving to, and further developing of this architecture. As the architecture has been developed with the business model of the network operator as the primary focus, these first sections discuss the benefits SDN brings to solving identified business problems.

A high-level overview of the networking components and how they can be used to benefit the business is given in Chapter 7, "An outline of the functional components."

"SDN: Defining a Strategic Business-Focused Architecture" is further broken down into its key component areas. Each component area has its own chapter:

SDN: Service and Infrastructure Architecture

Chapter 8 focuses on new network infrastructure technology solutions which become available through SDN and the other key technologies. Examples of the topics covered are SDN VPN or SDN Data Centre architectures.

SDN: Service and Infrastructure Management

Chapter 9 highlights the new management tools which are becoming available and identifies new tools which could be made available within an architectural toolbox. It discusses how these functional components can be structured to achieve automated and proactive service and infrastructure management.

SDN: Network-integrated OSS Architectures

With SDN fixing many of the networking limitations that have constrained NMS and OSS systems development, chapter 10 addresses the implications to the OSS stack, of the SDN network and policy controllers delivering automated and abstracted configuration into the end device. It also addresses the integration of bi-directional APIs into networking and identifies how a SDN architecture can utilise Big Data for service management, control and delivery.

End-to-end description: SDN Strategic Architecture

Chapter 11 brings these together, describing how the SDN Strategic Architecture could unify these three areas of components into a layer architectural structure.

The final chapters discuss how SDN Strategic Architecture could be evolved using an object-oriented approach and how it relates to working practises used within Agile development and DevOps control. Also discussed are the new evolving IP protocols of Network Service Header, IP Meta Data and Segment Routing architecture and their uses for service control and management.

1. Introduction

For some years now, the network operator industry has been going through an evolutionary period of major technology developments with the creation of Cloud technologies, Software-defined Networking (SDN), Network Functions Visualisation (NFV), Big Data and Next Generation Operations Support Systems (Next Generation OSS). These key innovative technologies target at reducing complexity and controlling service delivery for the network operator, and aim to bringing greater flexibility, agility and simplicity into the network operator environments.

The primary focus of these technology developments is to solve the business problems that are constraining the network operators, and to open up new revenue opportunities. These technologies are not created for the sake of having new technologies. Although they are not as silo-based as historic solutions, they still have the need to be architecturally integrated to create a flow through automated business-focused architecture.

As network operators achieve revenue by enabling their services over a network infrastructure, the activation, control and management of the service and infrastructure is primary to the business model they operate.

SDN sits between the higher-level applications/technologies and the network elements. SDN takes the generalised business logic and abstracts it into the network elements with the purpose of initialising the service on the network infrastructure. In addition, it enables the APIs from the network elements to communicate greater analytics, therefore ensuring that the higher-level applications can drive enhanced service management and control for the benefit of the customer. This is done with the aim of creating greater efficiencies in the operating model of the network operator.

With SDN being the enabler technology between the network infrastructure and the higher-level service and business logic applications, SDN can therefore be identified as being the technology which unifies these other innovative technologies for the driving forward of the network operators' business goals.

Many in the industry already believe that SDN is the future of networking. The author believes that SDN can achieve much more. This is based upon work done over the last nine years that has been targeted at creating a new architectural business approach.

The purpose of this book is to share information on how SDN can be used to bring together the multiple new technology developments and to suggest how a new architectural approach can be defined.

Change always impacts the entire organisation; it requires investment and creates work. This book describes a strategic end-to-end approach that allows network operators to kick off their evaluation of SDN by investigating available solutions that address key problem areas within their organisation. This way, companies when they have identified the real business drivers to change, can start implementing at a small scale and evolve the technology into their environments. Through this approach lessons can be learnt, and the companies can develop their business-focused technology strategy over time through experience. This allows for the strategic approach to be communicated throughout the organisation and for the value to be seen by the people within the organisation, therefore minimising disruption and impact to on-going day-to-day business. When benefits are recognised within the organisation, this indirectly helps teams to understand the need for change within their own domain.

This book is not definitive - but has been written to highlight how the evolving technologies can be brought together under a unified and flexible architectural approach.

As few companies operate the same business model or have the same targets and goals, the SDN Strategic Architecture aims to enable architects, solutions designers, developers, operations staff, product teams and engineers to identify how they can create a flexible architecture which is specific to the targets and goals of their individual company's business plan.

2. SDN: Connecting technology developments

The many new technologies of SDN, NFV, Next generation OSS, Big Data and Cloud that have entered the market over the last few years have successfully defined new approaches within each of their areas of focus. As these technologies have developed, lessons learnt in one area have crossed over and have influenced development in other technology sectors.

This crossover of technology features, from one technology sector to another, now allows for the creation of unified business-focused architecture.

The following is a synopsis of the development that took place. It begins with Cloud. Cloud used virtualisation of COTS hardware and created an orchestration layer to define a new systems-focused data centre architecture. Automation was achieved through use of portals to instantiate customer services, which were triggered through the orchestration layer into the server farms. The success of these developments created a focus on simplification and automation across the industry and in turn this evolution identified the automation constraints caused by the existing networking approach.

At the time, some concepts, which would later be used in control of SDN solutions, were being investigated by start-ups and universities. They were looking to overcome problems with modelling the scaling of the Internet, and how to test and validate new networking concepts and protocols. Put to use, the concepts that had been developed produced positive results in network automation, and this quickly gained traction, especially within Data Centre Cloud environments. It had achieved automation through its use of APIs from the IT sector and through its ability to signal via a network controller into the network infrastructure. With SDN and Cloud advancing in development, new use cases identified the constraints caused by fixed network appliances within an automated architecture. NFV aims to resolve these architectural problems.

NFV development centred on already developed Cloud technology concepts and set the goal of using virtualised COTS hardware for such appliances. This approach is complementary to SDN and Cloud, as NFV uses service chaining to link virtualised instances to create the end-to-end service structure between the NFV instances that have been initialised. As with Cloud and SDN, NFV will have an orchestration layer for the control and management of the NFV instances. This provides an interface for other components of the architecture to request NFV functions to be initiated into the infrastructure.

With SDN APIs now able to provide greater levels of analytics from network elements, the need for Big Data structures in the end-to-end management architecture becomes apparent, and the companies are required to implement a central repository of data on their services. This solution is a key success factor for delivering business management in the information age.

The advancements delivered by these technologies now give access to service and infrastructure management data as well as to the control mechanisms, which are used to deliver automation of configuration into the infrastructure using policy control. Access to detailed infrastructure data now removes some of the constraints that have limited the development of the OSS and NMS stacks for many years.

Through the availability of critical service information, this allows for the OSS architecture to be restructured with a singular focus of commercial service management and control.

With the SDN, NFV and Cloud orchestration layers now delivering the technology management, control, administration and activation of the service and infrastructure, the OSS can now focus on the commercial aspects of service management and control. It will no longer have to address the technology complexity and the constraints caused by having to program proprietary networking technologies and end devices. Technology configuration and end device management can now be handled using an API to the SDN, NFV and Cloud orchestration layers. With the focus of the OSS stack on the commercial aspects of service delivery, this drives an

16

emphasis on products, revenue creation opportunities and customer service delivery.

As SDN technologies deliver the ability to automatically enable the service in the network for the customer it becomes the linking technology between the live service and the higher-level applications and technologies.

All the other technologies are critical for the completeness of the service management. Therefore the aim of this book is the following:

- To discuss a business-focused architecture for network operators

- To highlight and suggest a strategic architectural methodology which can interconnect these key technologies

- To suggest how SDN fits within an end-to-end architecture and how it can be used to link the other technologies

- To share the thinking of the author on how a new architectural model can be combined to create a business-focused technology solution

- To highlight the benefits SDN brings to network operators

- To suggest new service and technology development approaches

- To highlight further advancements that are needed to solve additional network operator problems

- To create a flexible business architecture

- To generate discussion and investigation into how a business-focused technology architecture can be enhanced by object-oriented thinking

Summary

With these technologies having evolved almost side by side and having learned lessons and requirements from each other, they have produced a solution that naturally fits closely together. In a

17

network operator's environment these technologies tend to be run by separate teams.

This book identifies how these technologies can be brought together for the benefit of each of the network operator teams so that they complement and interconnect in a structured and layered approach.

The aim is to produce a business service solutions architecture that is flexible and can incorporate change.

As this book discusses a strategic architectural solution, it should be noted that, although many of the technologies discussed in this book are available in the market place, some of the technologies discussed are still in development or are suggestions for new products. To achieve full functionality, the network operator teams will be required to drive the creation of the capabilities.

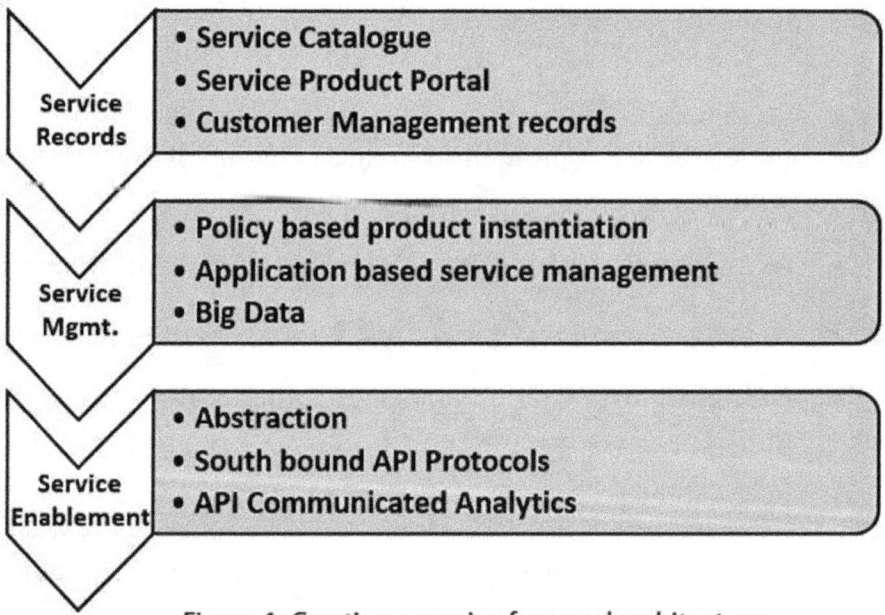

Figure 1: Creating a service focussed architecture

In addition, the open SDN architectural approach enables the network operator to use a more open model for sourcing of technology solutions. Through the use of the API capabilities used by SDN, the network operator can now integrate service management and control applications that can be developed in-house, supplied by vendors or developed through the co-operative

18

Open Source approach. This provides the network operator with the ability to create solutions that differentiate themselves in the market and to operate in a Web Platform business model.

As with all strategic solutions, evolution is necessary, along with the introduction of functional enhancements so as to achieve the goal.

3.   A Strategic Architecture

A strategic approach is required to plan the way forward for network operators. It is simply impossible to change the direction of these companies over night because of the complexity of technology, the existing systems, the business needs, staff training needs, regulation and the cost of operating and supporting such an organisation, etc. For a business-focused technology strategy to be defined it requires a greater clarity on how technology development is changing the landscape.

Therefore, with SDN evolving away from the hype curve and becoming a mainstream technology, the new developments SDN brings enables an opportunity to define a strategy that delivers a new approach to automation and network management within the IP networking domain.

'SDN: Defining a Strategic Business-focused Architecture' discusses how this new technology can be used to achieve an end-to-end solutions architecture which can be utilised for the development, expansion, operating and control of services and network infrastructures.

Even though SDN is still in development, it brings with it many tools, some of which have been developed within the Open Source methodology of functional development. This toolbox enables a layered architecture and uses open interfaces to bring together other evolving technologies, i.e. Cloud, NFV, Big Data and Next Generation OSS, and creates a unified end-to-end architecture.

Early adopters have already discovered the impacts that this cycle of technology development can bring, and they are educating and restructuring their organisations to gain the additional benefits of these technologies. Its impact has been significant, both for those who are already using it and for the vendors who have delivered the networking technology that has taken the Internet to where it is today. Operators and vendors have already identified that they can now deliver new products faster to market using a software-defined approach.

These changes provide the opportunity to deliver end-to-end, flow-based service control over IP networks. This differs significantly from the historic approach; when – with closed system routers with limited management analytics – organisations were restricted to managing the infrastructure devices as a means of emulating the management of the end-to-end IP flow-based service.

With the focus now on the management of the service, there are growing opportunities for new business models to be adopted, and the chance to create a new phase of innovation on the Internet.

3.1   Understanding the need for strategic change

Strategic change is defined as change in an organisation's business plan or marketing goals in order to achieve a new target. Due to increasing competition in the market, the senior management from many companies are now demanding a shift towards greater architectural flexibility, customer service and a faster time to market for their products. To make this happen, a new architectural approach is needed – one that better supports the other changes that are already taking place.

Additionally, there are many other business and technology reasons as to why a new end-to-end architectural approach is needed. The following list highlights limitations in the current approach that constrains companies from achieving their business goals. This list is created in no particular order and leaves considerable scope to add additional points.

- The current lack of unified end-to-end architecture drives proprietary solutions which increase workload and stress for staff across organisations

- Existing technology solutions provide few options to create automation or to reduce technology complexity

- Granular management and control of individual IP services flows is difficult to achieve. This is caused by the difficulties in isolating IP flows across infrastructures that are designed to aggregate.

- Maintaining the services and networks across multi-vendor solutions requires significant effort and skill

- Current component technology have long lead times for delivery of new features or scaling of equipment

- Current component-based architectures and solutions often have no common approach with regard to management and design

- Functional systems seem to get expanded to serve the competitive purposes of the vendor rather than being structured in clean functional blocks to better suit the architectural needs of the network operator

- The ability to extract and process granular analytics relating to an individual service is limited

- There is an on-going change from infrastructure-based services to flow-based services. Currently the infrastructure is managed in an attempt to emulate the management of a service, however with IP now being used to access a service from anywhere in the world, infrastructure management is no longer accurate enough to ensure the delivery of that IP flow-based service.

- IT systems tend to become more and more complex, and do not yet take full advantage of integrated Big Data solutions

- Offline IT systems are not informed about the real-time state of the network or real-time state of a service

- Offline IT systems cannot access the full analytics due to the limited output from the network elements

- The closed nature of optical systems and the problems of operating multiple vendor optical domains in one network operator environment

- An increasing number of standards which find little acceptance

- Technology development that doesn't deliver to the needs of the business, or doesn't deliver in a suitable time frame

- Roadmaps being driven by inputs from a few major operators

- Proprietary CLIs for configuration of end devices and network elements

- IT systems which are developed out of sync from the networking capabilities and which have limited service delivery control

- The cost of having to upgrade CPE, the impact to the consumer, the delay in time to market and the logistics costs

- Technology constraints which limit off-net business opportunities

- The need to significantly overbuild networks to achieve service and infrastructure redundancy

- Increasing level of complexity which requires highly skilled professionals to resolve even basic problems

- Too many scripts and quick fixes' in place instead of proper tools

- Too many projects get cancelled after significant investment

- Too many late nights doing out-of-hours changes

- The lack of insight to the problems of the business due to the inability to access analytics

- Too much repetition in development

- Invalid KPIs being measured

Any industry that is facing so many problems, all of which are understood by the professionals in the industry, needs a change of approach. This strategic change can be delivered by creating a cross-technology strategic architecture. Bringing together SDN, NFV, Cloud, Big Data and Next Generation OSS creates a unified technology approach that can be structured to meet the needs of the business.

Albert Einstein had it right when he said;

"You cannot solve a problem from the same consciousness that created it. You must learn to see the world anew."

When considering the statement above by Albert Einstein the following questions end up being asked:

- Why can routers not be automatically configured?

- Why is network management reactive and not proactive?

- Why is service management and control so complex?

- Why does network management not get the organisational, business and management focus it needs?

- Should routers not have a greater balance between bulk forwarding and flow-based forwarding?

- Why are real-time network modelling tools not integrated into the NMS to support the operations teams in decision making?

- Why do repetitive problems have to be repetitively analysed and fixed?

- Why does the OSS stack not get greater access to analytics from the network elements?

- Why is there so much duplication of development across projects?

- Why do we to install redundant routers?

- Why are there so many new IP protocols that don't seem to get used by the industry?

- Why can't an optical circuit be triggered upon demand?

- And many other questions...

3.2   Strategic Business Focus using Architecture

Complexity in networks and IT systems has now reached a point where it severely constrains how network operators can develop and deliver services. This complexity limits the ability of the network operator to deliver innovative and improved services to the customer in a cost-effective and timely manner. With so many network operators around the world being financially challenged, the demands to do more with less are increasing.

To reduce costs network operators are collapsing services onto unified backbones and with greater numbers of services moving to using IP, management and control of the flows that make up the service are becoming very difficult to identify, let alone manage.

Complexity also resides in the IT/OSS systems. Many of these are proprietary and operate in non-real-time. These systems have been designed to utilise the limited management output currently made available from the network elements. This has caused these systems to be functionally constrained and these constraints usually delay the development of new or enhanced services.

The limited management data has also constrained the operations teams who operate the network infrastructure and service management for the business. The management systems are generally reactive in nature and expect manual intervention. To overcome these constraints the business tends to offer SLAs that are of limited value to the residential customer or the business consumer and that, can lead to loss of revenue or customer dissatisfaction.

These constraints have in part been accidentally created by the current lack of an end-to-end architectural environment. This is impacted further as vendors add new features and functionality. This features and functionality creep ends up crossing over a multitude of functional responsibilities and over time creates structural architectural difficulties for the network operator. These multi-purpose functional component systems limit the network operator's ability to control their environment and tend to generate extra cost as the network operator has to purchase the full platform to access a sub-set of functional capabilities.

The continuing reduction in staff along with tighter CAPEX and OPEX controls limits the ability of network operators to release budget for new development opportunities or to address the existing operational limitations with technology. To address these challenges, network operators require a new business-focused strategic architecture that suits the individual business needs of the network operator and that can be delivered over time and within budget constraints. For these reasons, the SDN Strategic Architecture creates an architecture that unifies the historic CxO technology organisational structures, that defines a building block approach, and that sets expectations on the capabilities that components will be comprised of. This approach ensures the usability of the solutions and accessibility to the analytics.

Strategic drivers of the SDN Strategic Architecture

The strategic drivers for the SDN strategic architecture are:

- Enabling an automation infrastructure to link SDN, NFV, Big Data, Next Generation OSS and Cloud technologies

- Gathering analytics and pushing configurations for service and infrastructure activation through bi-directional APIs

- Providing building blocks based upon a toolbox

- Drive towards proactive service and infrastructure management

- Enable automated service control over access networks

- Unify access technology network control

- Incorporate Big Data into the operational service and infrastructure control

26

- Incorporate business security control into the OSS layer

- Focus the OSS on commercial service management

- Enable selected per flow service control for on-net and off-net traffic

- Enable the ISP to interoperate with the Web Platform companies

- Enable service flow management and service control

- Enable an infrastructure which delivers faster time to market

- Drive the ability to deliver accurate and internet wide customer service management

These drivers create a business focus on a fully managed service-aware network infrastructure which provides the ability for automated configuration and which has a focus on delivering new products and services for the benefit of the customer.

Strategic Agility of the SDN Strategic Architecture

For network operators competitiveness is somewhat restricted by the complexity of the technology and by the isolation of the technology components used within the creation, management and control of the infrastructure and services. This isolation creates complexity and in the process limits the ability of the business to stay competitive.

The SDN Strategic Architecture is structured to reduce the isolation of these components and to permit clean and clear interconnections where each technology component has a defined purpose and goal.

Examples of these tools include Big Data, NFV, SDN network controllers, SDN policy controller, LNMF modelling functions, a cognitive response engine, real-time resource management data base, real-time topology databases, vCPE solutions etc. Having these tools already in place and having their functionalities known by the designers avoids the duplication of functionalities during design processes. With having functional components defined, vendors then will focus on delivering the full functionality needed by the operator instead of trying to grab more market share by combining everything into one platform.

With developers operating in parallel, requests will be communicated to the relevant team that manages and owns the functional tool as to what data is required on the API. It is then the sole responsibility of the system owner to deliver the features and functionalities needed and to communicate the required data via the API to other parties. This ensures that, when creating new capabilities, the existing tools are utilised using APIs to deliver the new service, rather than creating a completely new set of functionalities because the original tool was missing a feature. With this approach, focus changes to fixing the system or replacing the system rather than bringing in additional systems that in turn functionally break the architecture. This forces the organisation to make the right decisions at the right time.

Strategic Goals of the SDN Strategic Architecture

The strategic goal of the proposed architecture is to provide a building block based architecture that brings together the various technology groups. This proposed architecture aims to enable the network operator staff to develop their products and solutions within an identified architectural structure.

The architecture aims to achieve this by utilising the toolbox of technologies provided by recent industry developments and by unifying these under a common structure. This does not limit the products that can be developed as the architecture is defined in a way that allows it to absorb new applications and processes. This permits those tasked with the responsibility of an area of development to focus on their solutions development with the knowledge of how they need to structure the output from their part of the solution through a pre-agreed API. This ensures a focus on the key output parameters at the outset of the development, along

28

with a clear understanding of what inputs they will receive from the other development teams.

Through use of the toolbox of applications and technologies defined in this architecture, the initial building blocks defined will enable the network operator staff to understand how to incorporate the tools into new developments. The approach is object orientated in nature and this concept is introduced later in the book. The aim is to reduce development times and in the process reduce the operational complexity of new products and services.

Achieving these strategic goals will drive a strategically aligned solution which will help the technical staff across the many divisions to jointly work towards the companies' objectives.

Strategic Plan within a SDN Strategic Architecture

This book discusses the strategic plan for the SDN Strategic Architecture. Its focus is the customer. This is because the end user generates the revenue, which, somewhere along the line, pays for the wages of all in the industry. The SDN Strategic Architecture considers the end customers' needs and brings together a technology structure that, when applied, is flexible enough to address the evolving and changing requirements of the customer'.

This focus on the customer is because the only purpose of a network is to deliver a service or data to a customer, in return for revenue or an opportunity for revenue.

Summary

The on-going evolutions of Cloud, SDN, NFV, Big Data and Next Generation OSS enable the ability to create a strategic unified approach. These evolving technologies focus on and enable the management and control of the complex technologies that the network operator industry uses. This change of focus enables the business focus to shift from infrastructure management, to service delivery. As these technologies complement each other and have been created to solve the problems that the network operator faces, this gives the network operator the opportunity to reconsider how they wish to achieve their business goals. Creating a strategic direction specific to themselves, as is permitted through this technology evolution, enables the network operator to move out of the current fire fighting mode and to set in motion a strategic direction that delivers an architecture that supports them in achieving their business goals.

4. Structuring a Service-focused Architecture

SDN discontinues the old model of integrated control and forwarding planes, which provides a new approach to satisfy the adapting business drivers. By utilising this approach the SDN Strategic Architecture has been created to enable technical staff to meet their yearly objectives by changing how they create the deliverables to satisfy the demands of the business. This opens up new approaches to create and integrate new tools, thus enabling the architecture to be more flexible, therefore achieving one of the most demanded managerial changes. New tools can now be created and integrated through Open Source initiatives, in conjunction with vendors or via in-house development.

This is achieved by using a layered model with well-defined handoffs between each of the technology development teams. The SDN Strategic Architecture is defined as a layered architecture that utilises defined and agreed APIs to interface between the different areas of responsibility.

SDN, Big Data, next generation OSS, enhanced BSS, SDN enabled NMS, NFV and Cloud, when aligned together, provide the necessary tools to address business needs. SDN facilitates the ability to interlink these technologies and to bring the business-focused architecture together. These changes drive the creation of new functional tools that focus on real-time network control and management and use automation of network configuration based on flow based programming, pre-defined templates or data configuration models.

Interconnection between the technologies is achieved through the orchestration layers that have been created in each of these technologies. When a capability is needed a call is made to relevant orchestration layer via APIs. This avoids end-to-end solutions having to be developed for each project and duplication of effort and workload.

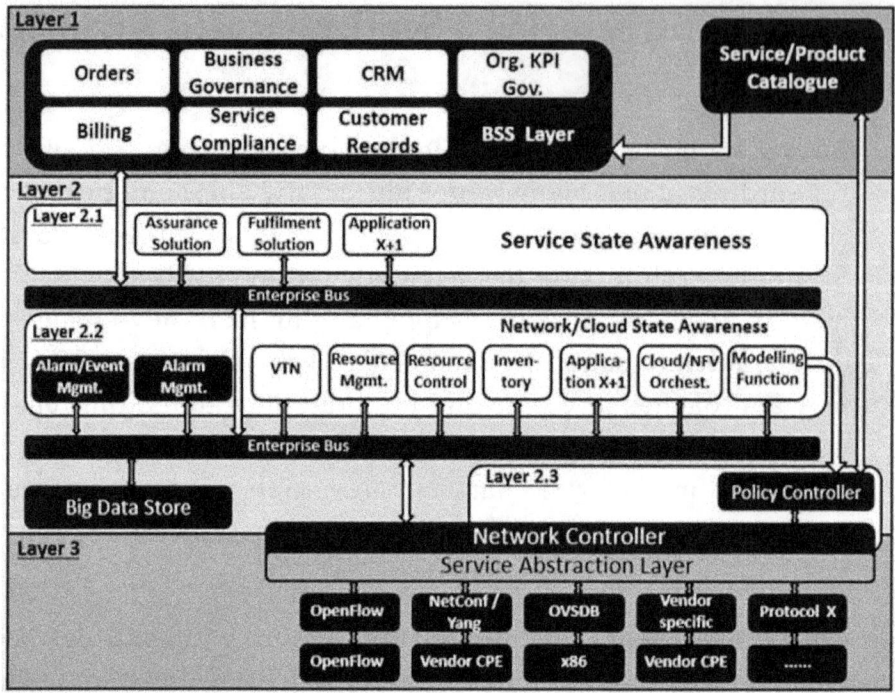

Figure 2: A overview of the layers in the SDN Strategic Architecture

This layered approach, created through its use of APIs, enables an architecture that allows for clear and precise hand-off points between the different development teams. These interconnection points enable parallel developments and reuse of already existing solutions to take place during project development. Separate development groups can now prepare their deliverables based upon pre-agreed API specifications which hook together the end-to-end deliverable. This meets the business needs of speeding up time to market by creating an environment that is agile and flexible. In turn, this enables a flexible and agile architecture that permits for functional expansion as new applications can be interfaced into the architecture through APIs.

SDN Strategic Architecture Objectives

The objectives of this framework are to target a focus on:

- Proactive fault identification

- Automation of fault resolution for service and infrastructure management

- Service delivery control through the ability to automatically monitor, manage and initiate services

- Delivering multi-layer technologies with the ability to self-learn, self-heal and self-expand, and which are end-to-end service aware

- Enabling policy based configuration control for service management

- Driving awareness of new protocols which expose service information to reduce the complexity of the network path

- Aggregation of service flows with the same service requirements

- Reducing the amount of repetitive tasks required to be completed by highly skilled staff

- Direct management of the service IP flows rather than service management through infrastructure analysis

- Simplification of network management

- Defining a new approach on how to deliver and apply security

- Use of real-time modelling to ensure the best utilisation of the network, depending on demands and trends

- Integration of subscriber service QoE for use by the ISP or multiple different ISP infrastructures

- Alignment of the strategies surrounding NFV, Next Generation OSS, Big Data, SDN and Cloud

- Enabling a higher level of serviceability for the consumer

- Real-time service assurance based upon service and user requirements rather than on networking per hop behaviours

As not everything can be done in a single 'big bang', the key to enabling such a change is to have a well-defined strategy that is clearly communicated and understood across the relevant departments.

5. Enabling the architects

SDN, NFV, Cloud, Big Data and Next Generation OSS technologies already create a significant change as to how architects, designers or engineers deliver projects. This chapter draws from twenty years of personal experiences and frustrations while working in the role of leading delivery of projects, and explains the reasoning for the bringing together of the SDN Strategic Architecture approach. It discusses how constraints can be removed and how those in the lead of projects can take back control. Key to this discussion is the reason why change is necessary to give ourselves the chance to be more successful in what we are tasked with delivering.

The ability for architects, engineers and designers to create a definitive vision for the future has been historically constrained by the component based technology solutions. This was caused in part by the organisational isolation of technology groups within their various areas of focus, the lack of openness in technology, the era of technology development and the inability to access the detailed service data from the network elements for accurate management of the end-to-end services.

These limitations caused an inability for cohesion across the groups and have led to a lack of unified focus on the end-to-end technology strategy, as each team dealt with fire-fighting within their own domain. This in turn has caused the emergence of isolated pockets of technology that struggle to meet the needs of the business or, more importantly, those of the customers. After all it is the customers who, by paying their bills, clicking on adverts, making a phone call, buying a video, downloading via mobile etc. have created the wealth for the entire industry.

This is the reason why this book focuses on defining an SDN Strategic Architecture with the customers in mind – and with the expectation that those in the lead will drive the continuing development of this approach as they apply new use cases to this model. This architectural approach is not definitive, therefore those leading this change now have the opportunity to put their own stamp on how such an architecture will be created.

Setting the scene

For many years the people in charge of the technical direction of "network operator" organisations (PTT, Web 2.0, Mobile, Cable, OTT and enterprise) have been in constant fire-fighting mode. Those in the lead have had to deal with rescaling their environments each year due to significant demand for capacity and functionality. This was required to be delivered while also consolidating networks, re-engineering old products, delivering new products and taking on new technologies. All this has been expected by the business within the constraints of reduced budgets and less resources.

These changes have not been minor. They have included product migrations such a TDM voice to VoIP, Analogue TV to IPTV, implementing TV products, upgrading to new access technologies, expanding data centres, incorporating new functionalities and capabilities into the OSS layer or managing layer upon layer of ageing technologies, and many others. This has had to be addressed while dealing with a growing number of security risks, regulatory issues and demands from management to achieve more with less. For most organisations this has given little time to fully evaluate the consequences of the direction that the end-to-end solution has been taking.

Few people in any organisation have had the chance or time to really ask the basic questions:

- Does the entire methodology we are moving forward with really meet the future needs of the business?

- Can the business and its people continue to absorb the growing complexity when the architectural environment does not include the flexibility and operational manageability that services and business require?

With change being a constant factor in technology, an integrated architectural strategy that focuses on both business and technology is required. Change can be delivered in two ways.

The first option for addressing change is to continue in the manner in which change has been historically addressed.

This approach has been addressed by the separate silos driving their latest thinking into the operator environment as they attempt to solve the very real problems for their group. In many cases the benefit and value that the individual silo uses to drive change, is mainly for the benefits of their own silo. These changes very often have consequences for other parties within the organisation, and these consequences are usually not fully understood or identified until long after the work has been progressed. In this environment change is usually perceived as a threat, bringing conflict and confusion.

The second option for addressing change starts with the acceptance of change firmly established in the thinking of those in the lead. As a consequence, they choose an architectural approach that can absorb change.

An SDN Strategic Architecture that combines SDN, NFV, Cloud, Big Data and Next Generation OSS is able to provide this capability. By using this toolbox, those in the lead can drive the direction of the end-to-end solution with all the vendors who supply the technology into their organisation. For this to be achieved, the decision-makers will need to broaden their perspective towards creating a flexible end-to-end architecture rather than only on the silo. Some of these primary enablers are listed below:

- Time to market for the product department

- Customisation for the product department

- Automated real-time management and control for the Operations teams

- Full visibility of services for the customer care departments

- A new supplier framework for the procurement department

- More clarity for the engineers, designers, architects, and operations staff

- An end-to-end solutions architecture toolbox to enable and simplify the effort required by their colleagues to deliver future proofed and unified solutions.

To enable change, the SDN Strategic Architecture discussed in this book utilises a layered toolbox of functionalities and APIs. This aims to permit new functionalities and systems to be incorporated into the end-to-end solution without major redesign.

Change required from partner vendors

When considering change, it is not just the network operators that need to change but also the vendors. Many vendors who understand and recognise the full impact of the coming changes have already initiated very significant restructuring both at a technology and organisation level within their own companies. This is clearly reflected by the recent acquisitions and product announcements that are discussed in the industry press. Vendors are now crossing over into other technology domains that have not been their historical focus with the aim of creating solutions-focused companies that are best positioned to service their customer base.

Proprietary equipment and systems have been a factor of networking equipment and OSS/NMS domains since the start of the industry - therefore it is this proprietary technology that has evolved the internet industry this far. SDN, Next Generation OSS, Cloud, Big Data and NFV impact and change this historic model. With the focus now moving to service management, this creates considerable space for vendors to be successful, while enabling a more open model from which the network operator - by committing to open source development - can redefine their options as to how they source their technology. This new open architectural option ensures that, going forward, the network operator staff are the guardians of the levels of openness that is required in the industry.

It has been the vendors with their expertise in scaling the network elements that delivered the technology that meets the needs of the operator environment today. However, the dominance of technology, protocols and features delivered within the appliance has driven an appliance technology focus, rather than a business and architectural focus for the development of many of these appliance-based solutions. This approach creates multi-functional component appliances that cause considerable complexity for the network operator when utilising and structuring them with their architecture.

In the past, many operators have had to wait for the vendors' roadmap to be announced before they could work out what products they could launch. This inability to firmly drive the future direction of a company has sometimes caused issues for businesses and for the people whose role it was to deliver products and operational solutions. As network technology has been appliance based, development of new solutions took years. When functionality limitations were encountered, these took some time to address and tended to impact the network operators' business.

SDNs separation of the control plane and forwarding plane, its focus on service and infrastructure management and its focus on software development aims to shorten timelines and to sharpen the focus on the end-to-end solution. This now requires the vendors who deliver singular components to enable an open and flexible architecture that supports open interfaces and programmable solutions based on open standards.

To fully realise the benefits of SDN, SDN needs to be considered within the building blocks of an E2E framework that incorporates NFV, Next Generation OSS, Cloud and Big Data. This is an opportunity for an SDN Strategic Architecture layered approach to be used to address the business needs, through providing the network operator with the opportunity to incorporate open source and internal developments into their E2E business solutions.

Roadmaps for many of the vendor supplied fixed appliance architectures were driven largely by inputs collected from a few major accounts. This happened because in many cases only these companies had the resources to fully investigate the future needs of their organisations. It can be argued that by accident the industry may have followed a direction that only a few had asked for and that this approach has led to the development of minimal variety on offer to customers. This in turn may have limited possibilities and may well have stifled competition as companies have had few options as to how to can utilise the available and scalable technology, to uniquely position themselves within the marketplace.

Creating business focus through architecture

The SDN Strategic Architecture permits for the developing of an end-to-end architecture that can be particular to an organisation. This is based upon the use of some or all of the common building blocks that are described within the SDN Strategic Architecture. Those in the lead from the network operator organisation can now select and enable building blocks that are particular to their business direction. This allows for greater choice at different price structures to customers and permits the operator to access technologies from a greater number of sources. Having access to more solutions sources will improve the creativity of the solutions that are made available, and it will drive greater variety of products into the market. This permits for the creation and use of customer-focused solutions that, within the architecture discussed, drive a path towards automation of fault resolution and the introduction of new application and features.

Cloud, SDN, Big Data, Next Generation OSS and NFV drive solutions that address the evolutionary problems that have occurred. This requires a reworking of some of the IT systems (OSS, NMS and BSS) functions. Historically these systems have been designed to cope with the limitations of the historic networking functions which shared little of the data relevant to the management of the service and which required programming in unique and proprietary CLI languages. These limitations have had an impact on the evolution of the OSS/BSS/NMS systems. The limited analytics available led to OSS/BSS/NMS systems being developed which attempted to evaluate what might be the state of the service, rather than being able to identify actual the state of the service.

These technologies bring a full focus on service management through new functional technologies that aim to enable automation of capacity and flow control. As an example: One of the targets of SDN is to remove the manual CLI configuration that is currently used to achieve network device programming. One approach is the use of instruction sets that can be triggered by using a policy controller. This change creates a building block that enables automation of service delivery.

Taken in conjunction with the Live Network Modelling Function (LNMF)1 , this permits for intelligent decision-making capabilities to be included into the network, which can enhance how an operator's services can be automatically delivered, managed and controlled.

Summary

Key within the SDN Strategic Architecture is the principle that change is expected, evolution is necessary and new services and requirements will appear as the shackles are taken of the service control, delivery and management. These changes provide those in the lead with the tools and the ability to tune the solution to the specific needs of the business model of their organisation. This allows a new approach that does not require those in the lead having to find a way of using solutions that were designed by other parties, suiting the business model of that other organisation.

---

[1] The term **Live Network Modelling function (LNMF)** is an author defined term to identify a network modelling tool. It gathers infrastructure and flow based data from the network for analysis. This system models the data gathered from the network to identify the network utilisation form the physical layer to a flow level. The aim of this tool is to analyse the network to identify trending problems and to drive the creation of proactive fault analysis capabilities

With this new level of flexibility and control comes greater responsibility and governance for the strategic technology direction selected for the company, by the architects, designers and engineers. However the flexibility introduced gives the ability to change direction as the focus of the business changes.

This architecture does need tuning and will require additional applications to be added to the toolbox using APIs to achieve the specific requirements of the individual network operators.

This is essential to creating a successful architectural approach, and on this basis the authors do not consider that the SDN Strategic Architecture will ever need completion - as completion is when evolution of the solution stops.

6. Background to SDN and NFV

This chapter casts some light on the background of the origins of these technologies and discusses why and how they evolved. In addition, the author offers his thoughts on some of the common questions that have been raised about these technologies, their direction and impact.

SDN as a concept was created with the goal of enabling a virtualised environment for research institutions to utilise live networks to test and validate new protocols and concepts. Its aim was to create the ability to emulate the complexity and scale of networking on full-scale live network environments. This came about because networking had become so critical to the world economy and because of the accelerating complexity of networking.

Quickly it was identified that the approach being considered could also be used in live environments, and that it could resolve many of the issues inherent in networking. Because it was software-focused, it was easier to adopt, although initially only in relatively limited environments.

As investigations continued, more expansive architectures and concepts were proposed from across the industry, and this shared thinking was incorporated into the solution. For networking infrastructure other parallel developments solved problems that SDN had not addressed. These included the development of the large scale LSR, the integration of the optical plane through GMPLS and the development of the real-time modelling functionality.

These parallel developments prepared the path for the introduction of SDN controllers as these can now leverage these technologies. This approach drives a new direction in the automation of network configuration, fault resolution and enables the inclusion of intelligent decision-making capabilities. The advanced development of these systems lends stability and allows for the focus to be put on creating proactive network management across multi-layer technologies.

One of the positions that have driven the SDN movement has been the closed nature of networking equipment. This occurred because of the integration of the control plane and the data plane within single appliances. This tight coupling of network control and forwarding plane creates a limitation for the operator in that it is difficult to access the management data generated by the appliance, this hinders operator and other vendors developments. SDN separates the control and forwarding within networking equipment and creates a point in the network that has a holistic view. This function then permits network operators to have an end-to-end view on what is taking place in their network. With the use of applications, end-to-end decisions can now be influenced through real-time off-line management and control systems which have an end-to-end view on the service and infrastructure. These systems and analysis add a level of intelligence into the decision making on how the network operates and is controlled.

6.1   Evolving SDN with NFV, Cloud & NG-OSS

SDN and NFV are some of the catalysts for the change that is now shaking up the industry. However some of the flexibility of SDN and NFV comes from the technology changes that were created in Cloud environments. NFV heavily utilises Cloud technology and infrastructures to deliver its capabilities.

Next Generation OSS solutions have been - and still are - developed by some existing but also new vendors. These new OSS solutions bring the thinking of SDN, Big Data, Cloud and NFV together, to realise a real-time solution for service delivery, control and management.

The solutions approach explained in SDN and the SDN Strategic Architecture unifies the technologies which have been created for Cloud and NFV and allows the control and management mechanism to bring the Next Generation OSS into a unified end-to-end architecture.

Whoever you speak to, be they a vendor, a new Open Source organisation, a start-up, a venture capitalist, an ISP or an analyst; they will all still have their own position and thoughts on what SDN is. Many see it as an opportunity to create a start-up and sell it off, a revolution, just another gimmick or just a data centre solution. However, as SDN has now moved of the hype curve, more and more operators see it as the building block upon which to deliver a unified architecture, one that allows them to operate their organisation effectively and to deliver, control and manage the services their customers want.

SDN, NFV, Cloud, Big Data and Next Generation OSS do not deliver all of the solution out of a box. What these combined technologies deliver is the ability to create a new architectural approach that enables advanced and simplified tools allowing for better and more complete solutions to be developed by the architects and design engineers. The tools sets that will be used will naturally be dependent on the requirements that the different network operator business models will require to best suit their individual business aims and goals. It is now the role of the architect within the network operator to identify the most appropriate method to structure the toolbox for the company to deliver the company's business goals.

The best starting point to explain SDN is to strip back what it is not and to remove the primary confusion between SDN and NFV. SDN is not the same as NFV. NFV addresses systems function virtualisation and deals with the virtualisation of various systems, such as firewalls or load balancers. For those who know about of Telephony Soft Switches, this approach is similar and could be described as an early attempt at NFV where dedicated appliances were replaced by COTS devices. What soft switches did to the now historic TDM switches, NFV will now do to other network functions. It will change these systems from fixed appliance-based hardware to flexible and agile systems that can exist and be easily spun up when required within a virtualised environment. This is an essential aim and goal for the creation of a flexible service-focused architecture and, although it significantly enables the goals of SDN, this is not SDN.

6.2 Network Function Virtualisation

Operators' networks consist of many equipment types (functions) that have historically been used to control a range of higher layer IP service functions. These service and flow management systems (functional systems) are used to achieve visibility of IP traffic flows, with the aim of increasing the management capabilities of the service or to secure traditional IP services.

Many examples of these functional systems exist, with the list growing longer each time a new capability is required. For the purpose of clarification a few examples are listed. These include Firewalls (FW), Network Address Translation (NAT), Threat Management System (TMS), Deep Packet Inspection (DPI), Load Balancer (LB), Intrusion Detection System (IDS) and Intrusion Prevention System (IPS).

These have historically been built for scale in dedicated hardware platforms that, due to their fixed platform architecture, have constrained network and service architectures because of their lack of flexibility in how they can be placed into the network. Although these higher function systems have constrained the flexibility of the architecture of the IP networks, they have permitted the delivery of services with greater security and control, and therefore they are not to be undervalued. Network Function Virtualisation (NFV) does not aim to remove these capabilities from the network; its aim is to deliver these functions in a virtualised environment on Common-Of-The-Shelf (COTS) hardware. The aim and purpose of this architectural change is to permit these functions to be flexibly activated in the network, therefore allowing for a more adaptable service-based architecture to be used by operators, thus enabling them to deliver a faster time to market for services and to address operational problems.

SDN and NFV will deliver very different capabilities into the network. Even though their deliverables are not directly aligned they are very complementary and together drive considerable control and management of the service for the end customer. By enabling the capability to spin up a virtualised function where required through the SDN and NFV orchestration layers, SDN can leverage of the capabilities that NFV functions. This can be achieved using the APIs that are becoming available in the NFV orchestration layers.

When NFV capabilities are considered as API driven control points in a service flow, this enables a new approach to service control: Service control that can be achieved through a flexible, scalable and real-time policy-controlled SDN Strategic Architecture. This capability also drives one of the key requirements in networking: End-to-end service management. Through APIs, service management data from the NFV applications can be extracted in the same way as any other application.

6.3   Frequently Asked Questions

APIs versus Signalling?

What is wanted from SDN Strategic Architecture is the ability to define and manage the services for users. One outstanding question is whether there are there too many APIs and whether some new signalling protocols could be used to simplify the approach of delivering the data between systems. Open rigid signalling protocols such as BGP brought the Internet to its current scale so we all know these work but are constrained. This structured approach achieved the Internet's expansion and such an approach should not be discounted as SDN moves forward for extremely common API creation. An example of a flexible protocol is the Network Services Header protocol that is going through draft at the moment of writing. This protocol defines the structure and leaves considerable flexibility in how the protocol is used and what data can be positioned into the protocol communication. This approach permits parties who use the protocol to define how they will use the data communicated. This approach is slightly different to the rigid standards of the past and could be something that is considered for some of the more simple API structures.

Is networking now IT?

Much IT systems terminology has entered the networking domain. Examples include the software focus, programmable APIs, policy controllers etc. Network engineers could be fooled into believing that networking as it was known is finished and that the IT department now runs the entire organisation. With this change there is no doubt that networking staff will need to learn some new skills. Learning new skills has been the standard experience within networking since it began and this should be of no issue. Template and data model driven creation should pose little problems to those who have had so many years of experience using CLI based programming. The need to have a deep understanding of the essential backbone protocols continues, and with that the role of the network engineer will be around for a very long time yet. Skills development programs that will provide the training necessary are already being put in place by some vendors. When moving from managing the infrastructure to managing the service, the network engineers' understanding of the protocols will be essential.

Is SDN only OpenFlow?

SDN as a network term seems to have existed before OpenFlow, but at that time it had two definitions as to its meaning. At that time it had a focus on network programmability. Since then, the work done with OpenFlow that was promoted by the network operator industry has made it synonymous within SDN. It is promoted and supported widely by many companies throughout the industry and could be described as a purist form of SDN as it was probably OpenFlow that put SDN on the hype curve. SDN from an industry understanding is now more than OpenFlow but the concepts that OpenFlow has brought has shaken up the industry in a very positive manner and has driven very significant change. Development is still needed before it can be used as the single control mechanism for the control of every flow on the network at points such as peering, IP backbone, optical, access and data centre etc. Whether those in charge want to develop it that far, and whether it needs to take up that challenge, only time will tell.

Solutions and protocols which are complementary to the model of separating control and forwarding are available and do deliver the capabilities of the SDN strategic architecture. Some of these will deliver migration strategies that will allow for a cost effective reuse of the equipment already invested in the field today.

Focusing on the customer rather than technology?

The key focus of the network operator should always be the customers with their extremely differing demands. It is the customers who require satisfaction as they pay for the industry to exist - through monthly ISP subscriptions, clicking on ads, being marketed to, buying products and services online, acquiring knowledge, acquiring information, paying for or watching TV series, and so forth.

It must also be noted that most of the networks that exist around the world react and deliver according to the customer who has so far not been made programmable.

The irregular nature of their choices and irregular time slot behaviours allow for a combination of concepts to inter-operate and to control the flow of data through the network according to the needs of the service they have chosen. The products that the consumer receives utilise technology but they are defined by business and regulatory rules.

How is a Next Generation OSS addressed?

SDN Strategic Architecture integrates the next generation of OSS systems functionalities into the real-time control of the network. This allows for business logic to be applied to the service and moves networking towards aggregated flow-based forwarding; it integrates network and services management, enables solutions which permits simple interworking between various parties, drives policy control, enables analytics capabilities to ensure control and management of the infrastructure and service. Achieving this is a key aspect of the definition of the long-term architecture for any business and its architects.

What is SDN not?

- SDN is not just OpenFlow. Within the industry SDN has goals that are greater than the ones of OpenFlow.

- SDN is not Network Programmability. Network Programmability is a part of SDN.

- SDN is not just end-to-end service management and control. SDN has a focus on network and service management through network programmability, abstraction and controllers.

- SDN is not Cloud. Cloud is a technology that SDN inter-operates with to deliver the end-to-end solution.

- SDN is not a Next Generation OSS. SDN brings in new capabilities that the Next Generation OSS requires in order to be able to react to services in real-time or to manage them in near real-time.

- SDN is not yet defined as being an object orientated, agile end-to-end solutions architecture. However it seems to be evolving in that direction.

Networks are already good quality, why change?

This is correct; networks today actually deliver reasonably good quality Internet to many subscribers. However the physical model used in network design is based upon a very significant level of overbuild to provide redundancy and resilience in the case of incidents occurring. Technology is used to create the redundancy, but it is only network over-build that makes this possible. This over-build comes at significant cost which many network operators are under pressure to reduce.

The equipment and geographical connectivity used in network creation is expensive to develop, purchase, to operate and to upgrade. It tends to have a short life span due to the continued significant growth in traffic load. The redundant devices and circuits also increase the complexity of operating the network, and this drives the increasing OPEX cost of running a network. In addition to these basic points, there are the many others already raised earlier in this book. These concern the issues of operating and delivering services that are not being managed correctly because of the limitations in existing networking technology that limit so many other aspects of the network operators' business. SDN and the SDN Strategic Architecture factor these into developments and the direction of the evolving solutions.

IP is by its definition an unreliable protocol. With an increasing number of impact-sensitive Internet services using the IP protocol, which are sensitive to packet loss, jitter and delay, etc., many network operators have no other choice but to significantly overbuild their networks to deliver quality to their customers. This approach, however, does not factor in how a best-effort Internet is expected to deliver quality. The decision about delivering quality connectivity is a decision for the network operator.

Many of the traditional network operators complain that the OTT companies are making excessive use of their infrastructures with their new services. The OTT companies, on the other hand, argue that they only utilise the service that the customer has already purchased and which the network operator thus has already been paid for. These are both valid arguments, but they don't create a solution that would benefit the future growth of the Internet, which both parties require. Nor does it address the needs of their customers.

SDN Strategic Architecture connects the technology structures (LNMF, dynamic optical (T-SDN), move towards carrier grade routers) to reduce the levels of overbuild. It also identifies the mechanisms that could be used to permit the customers to set their expectations on quality to other parties across the Internet, thus opening new business opportunities for all parties. This can help drive innovation, which also supports the goals of governments who are driving net neutrality legislation.

What will the regulatory impact be?

For many services the majority of the traffic will still follow the Best Effort Internet. However, as the SDN toolbox can now be used to identify selected queues, this enables the capability for all of the network operators to investigate and to create products which allow the customer to request quality on a per service flow basis, and to have this flow controlled beyond the ISPs network. This provides innovation possibilities for new start-ups and new business models for such companies.

The business logic behind Net Neutrality legislation

Net Neutrality legislation has been set into law in some countries. The government bodies that have driven Net Neutrality legislation have done so because they consider the Internet as a critical vehicle for future innovation. They see this future innovation as a means to create wealth for existing and future citizens of their countries and to provide employment for their people. With such major goals, governments deem it of great importance that ISPs do not have the capability to limit and constrain how such future innovations can be accessed over the Internet by people who wish to use a service of their choice.

Best Effort versus innovation

Net Neutrality defines the concept of Best Effort but it has no clear-cut technical definition. This leaves a very open expectation of what it could mean and leads to confusion in businesses around the world. However, as technology is rapidly changing, to define a technical definition for Best Effort could consequentially constrain the need for technology evolution - and this would benefit few organisations or individuals. This is a classic 'catch 22' situation.

For future start-ups to be successful with innovative services, the question that must be asked is whether only a Best Effort Internet can meet the needs of future innovative services, or whether technology can be used to measure the achievements of ISPs in achieving more defined and evolving Best Effort targets.

Existing technologies are limited at high scale to guarantee the delivery of services, but this does not mean that going-forward thinking or technology should remain that constrained. There is no reason to assume that customers should not be permitted to select and receive Quality-of-Experience on a service of their choice. This premise already exists in Net Neutrality ideals and in most legislation today. The consumer can request Quality-of-Experience (where such a capability is supported) for a service. The key reason for its lack of adoption is the inability to prove that Best Effort still exists when one user is prioritised over another, and the technology limitations that exist today in delivering Quality-of-Experience both on and off-net.

Assuming that one day there will be a flexible and evolving technology definition for the term "Best Effort", it will be required that technology is capable of creating a system that is capable of measuring and reporting on whether net neutrality target is achieved in a fair manner across the end-to-end network. SDN with its holistic network view and network management focus brings management tools into scope which could be used to answering these outstanding questions. Such a capability could be provided through additional developments of the LNMF (Live Network Modelling Function). With the LNMFs ability to model the network and the services, it should be feasible to identify the available capacity on the end-to-end network. With a network capable of being accurately modelled, this would then permit the network operator to "over provision" capacity within their network and to utilise and to control this capacity for priority services.

SDN off-net technology possibilities

Some key technologies in use today are ISP domain restricted and are currently not financially viable across the Internet. Through an SDN Strategic Architecture Quality-of-Experience can be achieved by permitting the end user to request Quality-of-Experience for selected services across ISP and OTT companies' networks. This can be achieved using signalling between network operators from their Command and Control systems and through use of IP Meta Data and NSH protocols. How this could be achieved is discussed in the chapter 16.

Additional Internet business models

A greater variety of business models are required to be supported on the Internet. The key focus of many companies is marketing through gathering as much personal user data as is feasible. This leads to excesses by some parties which breach privacy expectations of users and which can expose the Internet subscriber to personal security and privacy risks. Many parties argue that the consumer doesn't care - but there is an equally valid argument that many subscribers to the Internet are simply unaware of the extent of personal tracking. After all, not everyone who uses the Internet is a technologist.

In the authors opinion, privacy has not always been possible in society, however for him privacy is not the real focus of the discussion or for many others that the author has spoken to. The real focus is trust. In society all people have always had the right to choose their friends and to exclude those who they consider to have treated them badly. One of the negatives of the Internet is that 'bad friends', through their monitoring and data gathering, can remain anonymous and unwanted within the individuals "circle of friends", even when steps are taken to exclude them. Unfortunately, this breaks a key element of what has historically existed within the concept of privacy: Trust.

By providing the technology to create new business models, companies have different opportunities to use different means of revenue generation - and breaking the customers' trust can therefore remain untested.

Having put the customers and the services first, the points above can now be addressed in order to meet the customers' needs. To achieve this, automation and the enabling of service awareness will be required. This can be achieved through interaction with the customers' choices with regard to service quality, and the automation of the interconnection between the service and its requirements for network control.

Enhancing the architecture

Today's Net Neutrality legislation concentrates heavily on the physical connectivity but it can be assumed that it will adapt in the future to also focus on the logical connectivity to content.

Key questions for innovators and architects are: What could Open Source do in this space? And how can the architecture be enhanced through the growing focus on Content Centric Networking to address the identification of information and service, while providing openness and security for the consumer on the Internet?

7. An outline of the functional components

This chapter gives an overview of how the SDN infrastructure control and management systems can be used to create the toolbox for service and infrastructure control. Much of this book and this chapter are based on two papers that the author wrote and shared over the last five years with other operators and vendors. This book discusses the ongoing development of the ideas represented in those original papers and the reasoning for these new technologies and protocols. It introduces these components into the thinking of the reader in preparation for the topics covered in the following four chapters.

To control a network and the services running across it, SDN delivers a central point of management using the network controller. The controller creates an interface point for infrastructure and service management and provides a control point for the signalling of control using APIs. The following diagram identifies the key network systems and control points.

A Command and Control (C&C) system

This is an architectural concept system that comprises of a group of functional management components. SDN technologies through the network controller provide a centralised point to view the network. The C&C system takes advantage of this centralised view and creates a structured applications layer which utilises the APIs to gather data to create a holistic view of the networking connectivity, load, customer attachments/profile, service requests and location of service platforms etc. The C&C utilises the policy controller function to react to service requests and controls how the service requests are delivered through the network.

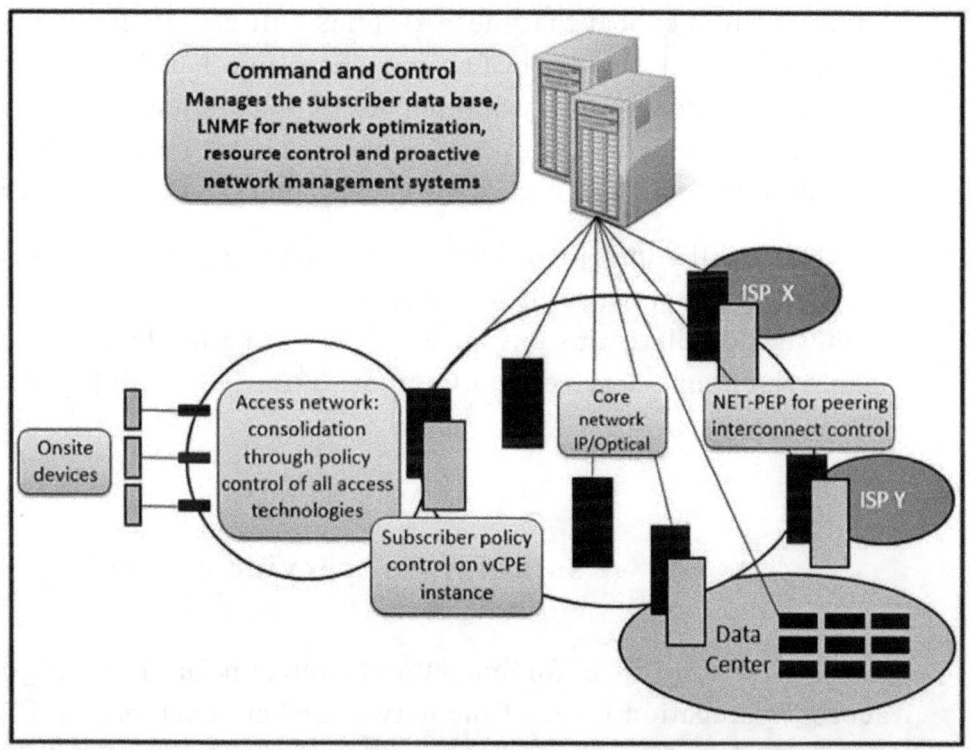

Figure 3: Command and Control Structure

A **subscriber policy** enforcement point can utilise technologies such as a virtualised CPE environment to apply control to the access of the subscriber. This control point retains the subscriber profile on the active end users who aggregate at this point in the network infrastructure. Local policies are applied to ensure their service requests are fulfilled. This ensures allocation of the appropriate resources to customers and ensures optimal networking usage in line with the quality expectations of the subscriber. Policies applied at this point interface to the pre-calculated paths on the backbone for the end-to-end control of the service and, as technologies progress, they will utilise NSH, IP Meta Data and SR to ensure service control. The logic applied at this policy enforcement point can be applied using a variety of technologies such as COTS vCPE, router vCPE, or routers.

A **network policy** enforcement point holds the service flow state and maintains the optimised path for the end user service based upon interaction of the C&C system. As technology evolution continues it is expect that these systems will utilise the segment routing, network service header protocols and that they will be configured through policy and network controller solutions. Examples of network policy enforcement point elements are Internet Peering Routers, Data Centre Servers/Routers, CDN equipment, etc.

The C&C will control and interact with physical or virtualised devices on the network. It will integrate towards the NFV and Cloud orchestration platforms to request services and to abstract management and service data to create a unified and integrated service platform.

7.1   A summary of service control

The PEP is a policy enforcement and control point at the edge of the access aggregation layer of the network where customer traffic is fed through and aggregated. This policy point can be a vCPE environment, and this system will have the ability to interface into the C&C policy/subscriber management systems. When the subscriber selects priority services, traffic will be directed onto the appropriate Segment Routed path across the core network infrastructure for that specific service.

For identified and supported services the C&C controller will have pre-computed the appropriate service routing paths using the Live Network Modelling Function (LNMF). This will ensure that the requested service will have the guaranteed capacity across the network and with the constant monitoring of the network new capacity can be added. The PEP can utilise the information in the Network Service Header protocol and its IP Meta data extensions to make forwarding decisions based upon the subscriber profile and products. This subscriber profile data will be extracted from the OSS/BSS systems (records can be kept locally using e.g. a LDAP structure) and will be used for defining actions based upon customer specified preference profiles. This permits for real-time service control based upon the customer preferences.

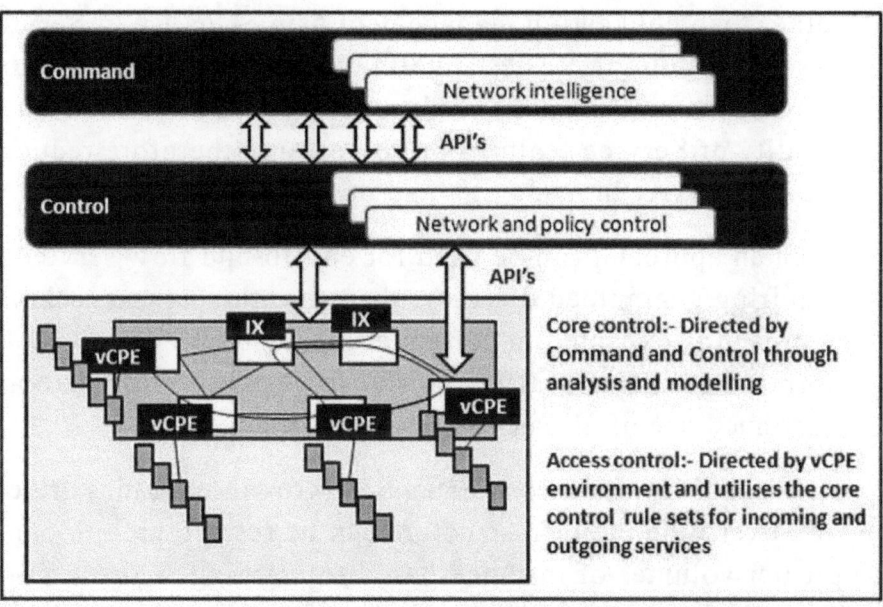

Figure 4: SDN-focussed Network Management

With the PEP having access to relevant subscriber information and with service information able to be shared across the network within the (IP) Meta Data fields from the service origination servers, the network can now be structured to react to customer service requirements.

This is further enhanced by the C&C having a full view of capacity utilisation, topology, latency and other relevant information to ensure the path selected meets the needs of the service and the consumer.

With Segment Routing permitting for path selection and aggregation of related flows with similar requirements, this permits the C&C to define only paths that have the capability to deliver the service as per the SLA. SLA information is obtained from the interface to the customer and product profiles.

As the C&C function will have the capability to signal new optical capacity or new Segment Routed forwarding paths, this allows for live network issues to be automatically resolved.

This capability to monitor network faults and current network state can then drive congestion avoidance to ensure service delivery, rather than congestion management using Quality of Service ASIC-based technologies. These evolving technologies and protocols provide the mechanism to reduce the reliance on ASIC-based Quality of Service features on routers and therefore reduce a lot of the complexity in managing and upgrading networks.

Such an approach can be used for end-to-end flow service traffic requiring guaranteed delivery; therefore the current technology approach is expected to continue to be used for the control of normal Best Effort traffic. This ensures a migration path and continued use of already sunken investment.

As technology evolves and capacity growth upgrades are carried out, over time the infrastructure can be restructured to cope with the greater volumes of instance-based forwarding. This permits for a migration strategy that enables the lifting out of specific flows from the network. From a business perspective, this avoids the need for a network rebuild and allows the introduction of new technology based upon normal upgrade and expansion capacity rules which all networks introduce on a regular basis.

The C&C concept has multiple different functional systems within its build. The Live Network Modelling function component can analyse and define the forwarding rules for flows, and through this it can include business rules defined by the network operator. These could include time of day, time of week and other variations, and these can be trended in accordance with expectations gathered from recent historical data.

These updates and the regularity of the tuning of optimisations would be dependent on the needs of the network operator's customer base. If for example the operator's network is mostly used during the day to support services for enterprises and during the evening or weekend for residential services, then appropriate optimisations could be modelled, tuned and pushed into the network to satisfy the differing requirements and needs of both business customers and residential consumers.

SDN Dynamic Network Control

Unlike the current fully dynamic routing protocols based solutions, the C&C LNMF will use real-time management data taken from integrated systems to react to situations occurring on the network. It also aims to have the ability to use historical data and trends to predict where capacity will be required based upon trend statistics.

By using the C&C online Live Network Modelling Function (LNMF), this permits for a programmable intelligence to be introduced to support the operating of the network. Through the SDN controllers the signalling of these "intelligent decisions" can be used to bring up more capacity, e.g. based upon knowing the historical trends, the current user needs, the services that are being used by the customers, identification of circuits starting to degrade or fail , etc. This allows for pre-optimised routing to be signalled into the network when new optical or segment routed paths are required. This capability allows to the network management strategy to move from being reactive to being a proactive.

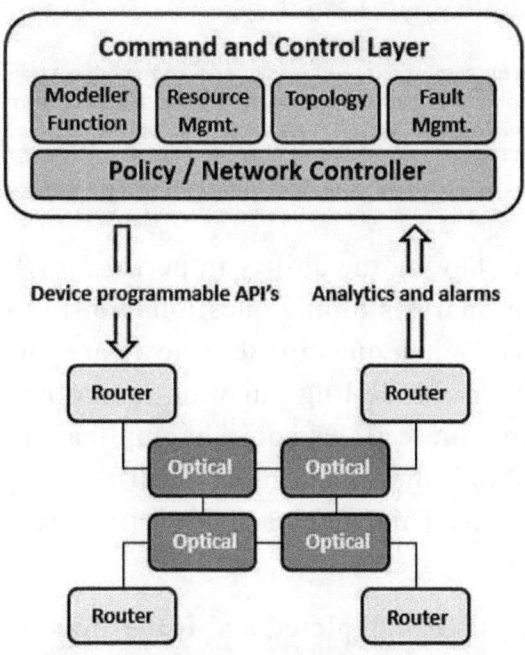

Figure 5: Command and Control focussed SDN Network Management

Aggregation of service flows using Segment Routing on an aggregated and managed path, ensures that the business logic of the service and customer requirements are delivered as the traffic passes through the core of the network. This pre-planning minimises the need for buffering and queuing because expected priority traffic can be scheduled based upon known and considered historical trends.

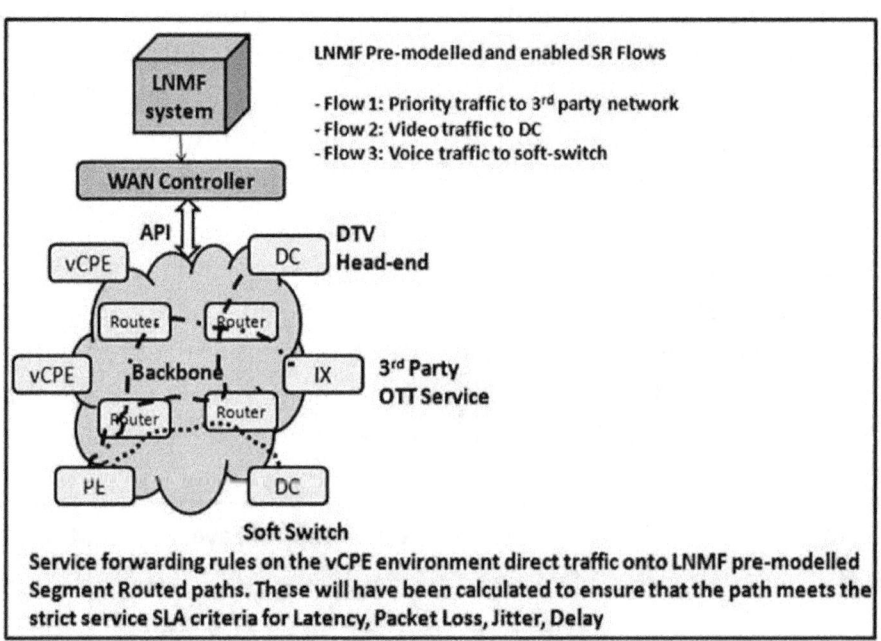

Figure 6: Pre-modelled Segment Routed paths

In addition, having the ability to perform offline analysis of traffic, permits for analysis of any questionable flows being forwarded by the network. When questionable flows are identified by the LNMF (Live Network Modelling Function) and other C&C management applications, these flows could be duplicated on the network equipment and forwarded to the security analysis system for in-depth analysis. This enables a simplified architecture for Threat Management Systems.

When analysis is completed and if the flow is identified as being a threat (DDOS, SPAM originator, hacking attempt), the instance can be black-holed using a forwarding plane rule instruction set at its first entry point into the network. This allows for flow-based black holing rather than bulk forwarding based black holing of threats. If the originating ISP/OTT Company has evolved their network and uses a C&C function in their network, a signal could be sent to the originating ISPs C&C, requesting that they kill the attack at source, therefore relieving latter ISPs from having to deal with or receive flows which are a security risk.

This concept introduces and creates the possibility for greater security on the Internet by permitting Security teams from network operators to interwork and to eliminate security threats.

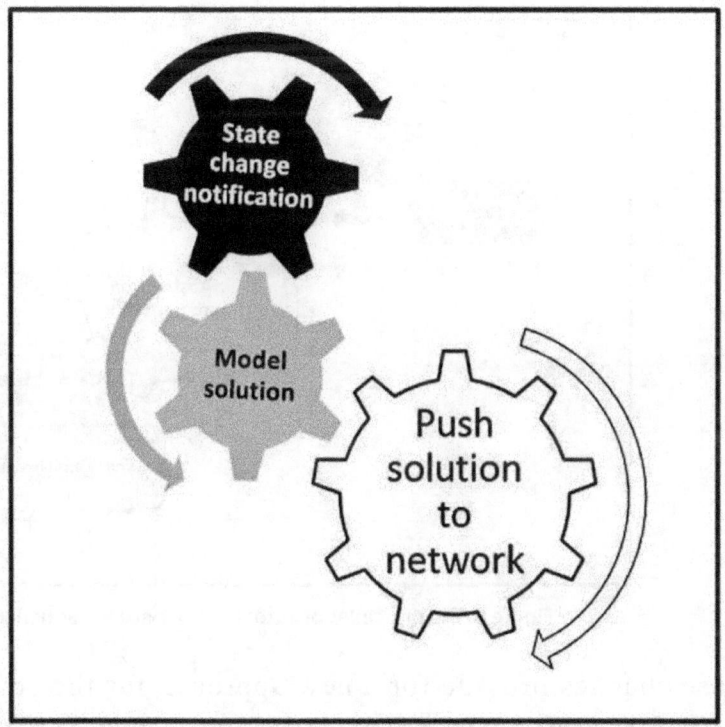

Figure 7: The principles of incident automation

When a consumer initiates a service, the initial flow can be identified within the PEP/vCPE solution. This approach permits for interactions to occur through the integration of real-time OSS/IT systems to allow for business logic to be introduced and used as a security tool. An example of such a security approach would be to validate that the consumer has purchased the service they are attempting to activate. If they haven't, there is no reason to permit the session to establish, as this would only require the Data Centre or CDN to process unnecessary traffic. This will be further

63

improved through interaction of the C&C and policy enforcement points using the new IP protocols currently going through standardisation. These protocols are the Network Service Header and IP Meta Data. These will make visible new levels of relevant information into the control mechanisms of the service flows. The C&C path selection decisions for service forwarding will be computed based upon awareness of the live network conditions.

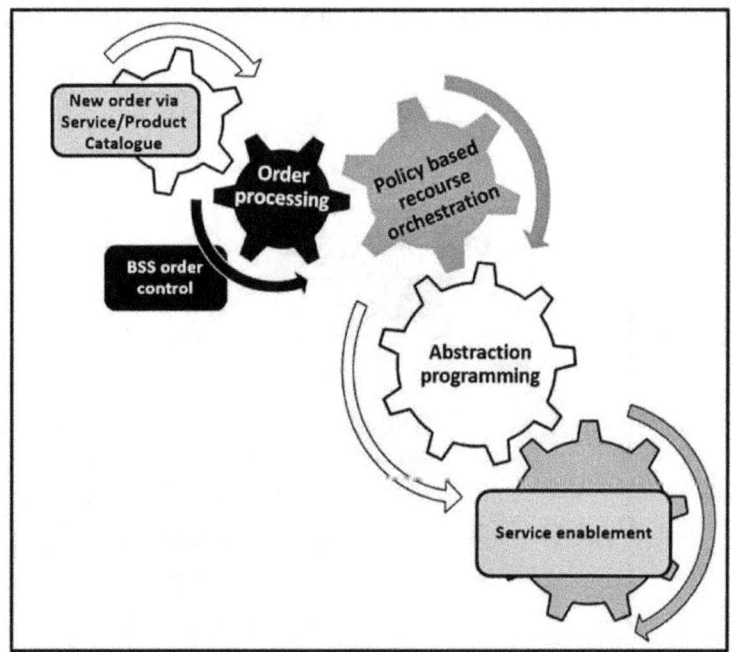

Figure 8: The principles of automated service instantiation

These changes provide for a new approach for the construction of traffic forwarding equipment, as it gives a focus on enhancing forwarding capabilities rather than protocol handling.

The separation of the forwarding and the control plane allows for the creation of carrier grade IP equipment, and with that a further reduction in the current network equipment architectural model in operator networks. Currently the architectural build model depends on duplication of equipment in the network. By reducing the complexity of the forwarding table creation through protocols within the network elements, the complexity of the active network elements is reduced as this leads to greater uptime and therefore less duplication of equipment for redundancy purposes in the network.

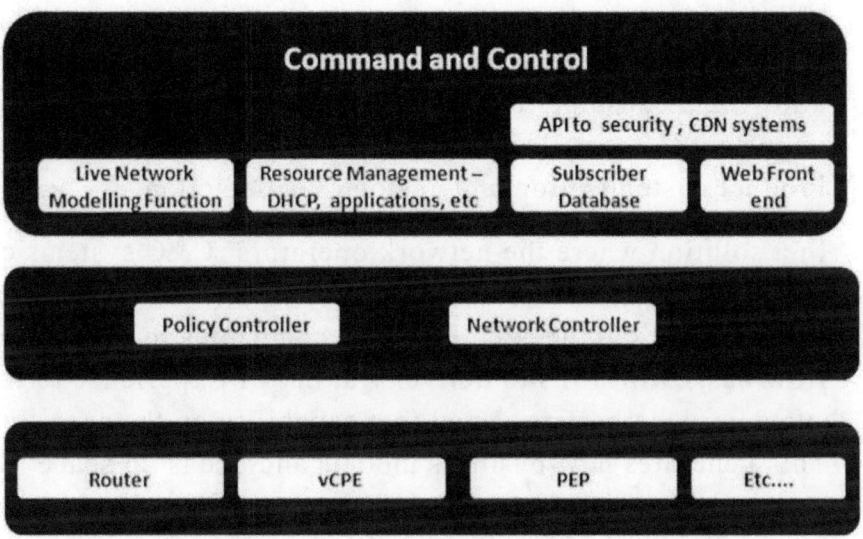

Figure 9: Layered management and control systems

Traffic flows change slowly over time. This mainly occurs when new peering points are put in place, new link capacity is added or because of manual intervention at a BGP level. As these additions are not done in real-time, the C&C LNMF function can both be used to plan and then optimise the network. The C&C can identify the optimal build for the network and the Policy and SDN controllers can trigger the appropriate changes into the infrastructure. For optical path restructuring, the LNMF will compute the models for optimisations and these can be activated using the controllers and GMPLS into the optical infrastructure. Currently work is underway on delivering an SDN (T-SDN) optical function, and with time this may further simplify the IP and optical integration.

Through evolving towards an SDN Strategic Architecture using the building blocks that have been identified, this will ensure customers receive Quality of Experience for the services they have purchased. This creates new service opportunities for network operators, and in turn generates new technology development opportunities for the vendors.

Due to its clear customer focus, this framework allows customers to purchase services from the ISP and OTT providers within a flexible model where decisions made by the consumer identify a variable model of payment. If the consumer wants all or none of the guarantees of quality, privacy, security etc., these could be factored into a product-pricing model by the service provider, therefore allowing a consumer to make their own decisions on what is important to them. This supports two of the main business drivers: Product customisation and customer satisfaction.

In a solution where the network operators' C&C systems can signal securely, new business model opportunities emerge. Such a model permits for example Web Platform companies to interact on a per-flow basis with ISPs to deliver Quality of Experience to the end customer who has requested that capability on the specific service. This generates new business models and opens up space for new start-ups to enter the market. This creates the potential to drive innovation on the Internet and allows the consumers to gain access to new services and opportunities.

7.2   Building block principles

When defining an architectural approach, multiple roads can be considered. The SDN Strategic Architecture defines an approach based upon a toolbox of building blocks. These building blocks provide tools (functions) that can be utilised to deliver the end-to-end architecture. Some of the network specific tools are discussed below through the use of examples.

Aggregated Flows

The Segment Routing (SR) protocol is an important new development for networking. It can be used to efficiently aggregate and transport bi-directional flows on a service basis for both IPv6 and MPLS. MPLS SR addresses the control of IPv4 flows across the core of the network. Native IPv6 Segment Routing delivers SR across the core of a network without the need for a underlying MPLS infrastructure. This has the potential to avoid some of the domain restrictions that occur with MPLS, as it could be developed to naturally extend pre-authenticated flows into another network operators' domain of control. In addition, there is no reason as to why over time NSH and IP Meta Data cannot be extended into the access networks, as this will provide the extra service information to the access technology networks to permit an extra layer of control within the access layer.

As the PEP can make use of the customer service business logic, this permits it to utilise SR to define on which flow the traffic will be forwarded. The business logic defines the expectations of the subscriber and how their service will be handled on a granular basis, e.g. all flows generated on the Internet connection to the OTT TV provider and identified web gaming systems will be priority, whereas all standard web browsing will be delivered at Best Effort. This identifies the expectations of the customer and delivers what the customer has paid for. With this knowledge, the PEP can now define which of the pre-calculated SR paths is appropriate for the service to be forwarded on. The LNMF will have pre-calculated the appropriate path through the network to match the needs of the service based upon the service requirements, and will have taken into consideration issues such as latency, packet loss, jitter, delay, and others.

Segment Routed capabilities will be further enhanced through the additional service information made available within the new protocols of NSH and IP Meta Data. These drafts, which are currently in the IETF, will enable the Network Service Header and IP Meta Data to allow for faster and simpler methods to identify traffic flows. This ability to carry service information will permit it to be easily forwarded into an appropriate Segment Routed flow when and where required. As NSH includes the control mechanism for service chaining, these protocols bring together the exposure of the service data, the ability to chain the service between the NFV instances required and the selection of a pre-modelled service specific SR path, which has been created to meet the customer needs based upon their subscriber profile.

CDN I/Federations

CDN-I/Federations allow network operators to inter-operate for delivering content to each other's subscribers. By using SR, IP Meta Data and NSH, rules can be pre-populated into the forwarding table on the router via the C&C with an end to-end view, rather than solely relying on the device-by-device view of routing protocols. This way, the principle upon which rules are set is business logic and network state rather than technology complexity - this ensures accurate measurements and guaranteed localised delivery of traffic. This provides the capability to ensure the Quality of Experience the customer has paid for and not a reduced variable bit rate stream. In addition where network congestion occurs because of high localised content demand at a single CDN replication node, the LNMF could be enhanced to be content aware. With the LNMF having a full view on the network infrastructure utilisation it could trigger to the CDN to spin up multiple deeper vCDN replication nodes, form where to stream high use content.

Data Centre/CDN Security

SDN in the Data Centre is already widely deployed and well documented by many companies. However, SDN brings additional value on top of the benefits that are already realised. This additional value can include scaling security at the Data Centre edge, as this is a significant overhead today. The C&C introduces an alternative through its end-to-end services-based approach. This capability is significantly enhanced because of the work done in the NFV, which permits for the accurate scaling and simplified deployment of network functions when required.

Enhancements to this can be made through having the subscriber policy enforcement point (at the vCPE) certify that the end user is valid, that the service has been paid for and that the service origin is correct. Having the flow already certified using business logic before it is sent on a SR tunnel to the Data Centre, reduces the need for highly scaled security equipment at a Data Centre/CDN edge for the serving of local customer traffic. This will require that the traffic flow be placed on a Segment Routed (SR) instance only when all the business logic checks have been validated by the PEP. This permits for a model to be developed where business logic can be used to validate the connection and not just IP flow. This can drive an alternative evolution for security and helps to reduce the complexity of the per-application security controls currently created by many operators. The last device (a LSR) in the network before the CDN/Data Centre can be configured on some links to not do bulk forwarding, but to only forward specific SR flows. With the ability to program the FIB to a granular level using SR, this could reduce the need for every flow to have to be checked using firewalling equipment.

Consider the following example: 60% of an operators Data Centre capacity is used to deliver local services to its own local customers (TV, VOD, email, gaming, etc.). Instead of using only bulk forwarding based routing into the Data Centre, segment routing would also be enabled. This enables separation and forwarding control of local customer service traffic from all the other traffic using the core network. This permits for the steering of traffic onto circuits that have been set aside for use by predefined Segment Routed flows. Services initiated by the customer through a vCPE environment can then be aggregated onto specific segment routed flows at the vCPE and forwarded across the core network to specific zones within the Data Centre. All flows towards the Data Centre service by the customer will first be validated based upon the products the customer has purchased. If the customer attempts to initiate a service that they have not purchased, the service will not start, as the customer does not have the entitlements to use the service. Through incorporating customer product policy within the vCPE environment this ensures the inclusion of business logic as a security check for traffic accessing the Data Centre. Through using a combination of business logic, the use of Segment Routing for the forwarding of flows and the use of dedicated circuits for the Segment Routed flows into the Data Centre, this reduces the amount of security appliances/instances required in a Data Centre. If an additional virtualised load control application is added to C&C function, this could ensure that the traffic is forwarded to a server that has free resources and that has the correct service application and content. This simplifies the firewall and load balancer build on the path and contributes to an easier scaling of the infrastructure. With such architecture, reduced traffic cleaning will be required in front of the Data Centre/CDN for these circuits.

IDS/IPS would be required, and if anything untoward was found to be occurring, an appropriately scaled NFV instance of a firewall could be spun up to address such issues. Any questionable flow would be redirected through a security analysis system, therefore allowing for a deep analysis of the situation.

Network Security

With the C&C having a full view of the traffic flowing through the network and having real-time capabilities to analyse flows using specialised C&C applications, attacks can be addressed more efficiently. If the traffic analysis system identifies an anomalous flow or irregular flow patterns, the first forwarding device in the path will be signalled to replicate the flow (e.g. like port mirroring, but on a flow basis) to the integrated C&C security solution for analysis.

A component of the C&C will be a security instance analysis system. If the analysis shows the flow to be a hack attempt, SPAM origination, part of a BotNet attack, DDOS etc., the C&C/PEP can enforce a predetermined action for the security violation along with tracking its origin and originator.

With flow-based security controls and the validation of customer service information, the ability for attacks and theft of ISP/OTT services etc. can be constrained. This requires additional development but fits within the building blocks identified for the SDN Strategic architecture. It provides new opportunities to control security as the architecture evolves, and as new applications are developed and existing systems enhanced.

Flow-based control between different organisations

With a C&C system in place, traffic traversing between ISPs can be constructed to communicate business logic, therefore delivering Quality of Experience for services going over different network operator environments.

Where service requirements demand extra service parameter information to traverses across network operator boundaries, the C&C could be used to set up secure connections by signalling between the network operators. This would allow both networks to enable the controls that are needed to ensure accurate control of the flows passing across their infrastructure boundaries. This enhanced NNI functionality may be used to service any traffic flow and would permit for the replacement of MPLS VPNs with SDN VPNs. This drives solid business models and enables flexible off-net Quality of Experience delivery mechanisms.

The following example could be used or developed to ensure end-to-end service understanding. When companies choose to work together, they could permit the network operators C&C systems to signal and assign flows using NSH/IP Meta Data. Additionally, the C&C systems in use in both parties' networks could signal parameters using off-line connectivity for additional security. These parameters can then be used to ensure a very high level of security, thus ensuring the Command and Control system can validate to a very granular level that the traffic flow is according to the company's business specific agreements. Individual flows can be extracted from the bulk forwarding traffic and granularly controlled for forwarding, choice of peering interconnects, etc. When the technology is aligned between different network operators, the customers' preferences can then be controlled according to the business needs.

Additionally, CDN-I capabilities can be enhanced, thus permitting for an efficient mechanism to localise traffic and to best guarantee quality to the end subscriber. This off course still puts the full load on the access network connection to the subscriber.

C&C – Integrated Management

The C&C differs from the SDN controller in SDN as it will involve not just the controller, but also the off-line command decision-making function. The command decision-making function integrates new network management tools and permits these to be invoked to react to situations on the network.

The command aspect of the system instructs the network to take action, based upon information gathered through the C&C application functions. Examples of these instructions can be: triggering additional capacity on specific paths, taking paths or wavelengths in or out of service, initiating Segment Routed flows into the network, and many others. These instruction sets will be intelligently modelled via the Live Network Modelling Function (LNMF) and then configured into the network using the SDN controller.

This will be based upon real-time data inputs gathered by the C&C management system, and after processing near real-time responses can be created. The C&C network management collector function is similar to existing systems included in NOCs all over the world today. However, within the structure of the Command and Control systems architecture the output from these systems will then be collated and inputted as scenarios into the Live Network Modelling Function for computational evaluation. Once correlation, analysis and faults diagnosis have taken place the reprogramming of the network can be using the SDN Policy Controller/Network Controller.

Having the Live Network Modelling Function advise on the invoking of the required changes allows for informed premeditated decisions to be made about the network. This enables automated intervention into the network to address network issues based upon C&C NMS data received from the network. A full live topology and inventory of optical and forwarding plane equipment will be required to be gathered and recorded, and will be used to trigger changes to the appropriate points in the network.

With the LNMF having a complete map on latency over time across links and paths on the network, pre-modelled options can be made available and can be pushed into the network, when required.

Some examples:

- The current utilisation of paths

- The ability to use GMPLS to trigger alternative paths or to identify paths with less errors

- The ability to do an analysis on flow anomalies and to investigate specific flows to understand security risks and to negate their effect

- The ability to accept Quality of Experience requests from 3rd parties based upon business logic

- The ability to gather, record, analyse and to factor in historical and up-to-date information about the network

- The ability to analyse latency, packet loss, delay, jitter and to set thresholds for each aggregated flow. This would allow an alternative path to be triggered when the current path no longer meets the needs of the service. This could be triggered on the optical using GMPLS or the IP forwarding layers using SR. This provides for a new level of business logic to be auto-instantiated into the network and used for control of service delivery to drive customer satisfaction.

The SDN Strategic Architecture also provides the capability to enable IT/OSS simplification. This is possible because of the SDN controllers' ability to abstract and trigger new configurations into the network. With the aid of the SDN policy controller, this allows for abstracted provisioning to the end device. When fully developed, this replaces some of the current IT/OSS systems functions and changes their focus to service control and management rather than technology control and management. This can be achieved in real-time and permits for a programmable approach. Additionally, the APIs provide the ability to gather data analytics from networked devices about the services and the infrastructure.

OSS and BSS business logic use APIs to communicate generic instruction sets to the policy and network controllers, where the configuration is abstracted onto the end device or the network element. The SDN controller facilitates abstraction at the controller level, and the southbound API delivers the programmability of the end device or end network element using a locally relevant and specific instruction set.

With the networking environment having now addressed many of its limitations the focus of the IT/OSS stack now changes. Now, rather than dealing with services at a network configuration control level, the IT/OSS stack could now focus on service control and management at a business logic level.

Forwarding Plane Complexity and Technology Reduction

A significant advantage of moving away from network device integrated control protocols, which have been developed in the technology lead networking solution era, is that some of these complex protocols can be reduced in importance. These could include IGP's, MPLS, LDP, RSVP, MPLS services (VPNs both layer 2 and layer 3), QoS, CoPP, and others. The majority of the complexity of network management comes from trying to understand the implications of these protocols; therefore their reduction simplifies network and service management.

Most of the protocols that are currently used forward packets based upon per-networking element based decisions and not on the goals of the end-to-end service and the business logic that is required to be considered. The SDN Strategic Architecture therefore simplifies networking as it enables a strategy that no longer requires individual devices in the network to only gain an understanding of the network only from constrained protocols. With its centralised view, which focuses on the end-to-end service delivery, a new and more simplified protocol suite can evolve that include the lessons learned from existing protocols.

While the desire is to reduce complexity, the SDN Strategic Architecture is complementary with evolving the existing installed base to enable a more efficient solution, with the goal of becoming more dynamic and service aware.

Access Network Policy Control

Intelligent bandwidth reservation on the access networks is already available with technology developed by the Cable, Mobile, and Fixed Line industry. This is known as PCMM in Cable, 3GPP in mobile, WMM in WIFI and OMCI in GPON. These mechanisms use solutions based upon knowledge of the customer profile, a policy server and application servers that are already partially aware of the current state of the access network. These systems signal into the access network to control the flow of traffic across the access infrastructure. Application servers can be developed and integrated with SDN controllers to further enhance the control of the access network and the service being delivered to the customer.

Having a PEP (e.g. vCPE) that is subscriber profile aware and can receive the IP Meta Data information contained within the Network Service Header provides the basic information to identify the needed capacity for the service requested by the customer.

Although many technologies exist, there is an opportunity to simplify the technology and to not carry forward all the complexity of the past. This can be made possible by integrating the SDN controller and abstracting instruction sets to control the different access technologies using a singular management system. Such an approach could create a more unified approach to access network management and control. Work already underway with Self-Organising Network (SON) could be considered for other technologies beyond RAN networks for improving the control of a multitude of access network technologies.

Through gathering of infrastructure analytics from the access networks it would be possible for decisions to be made as to whether the access network had the ability to deliver the service, based upon information obtained on the service from the subscriber profile and the IP Meta Data.

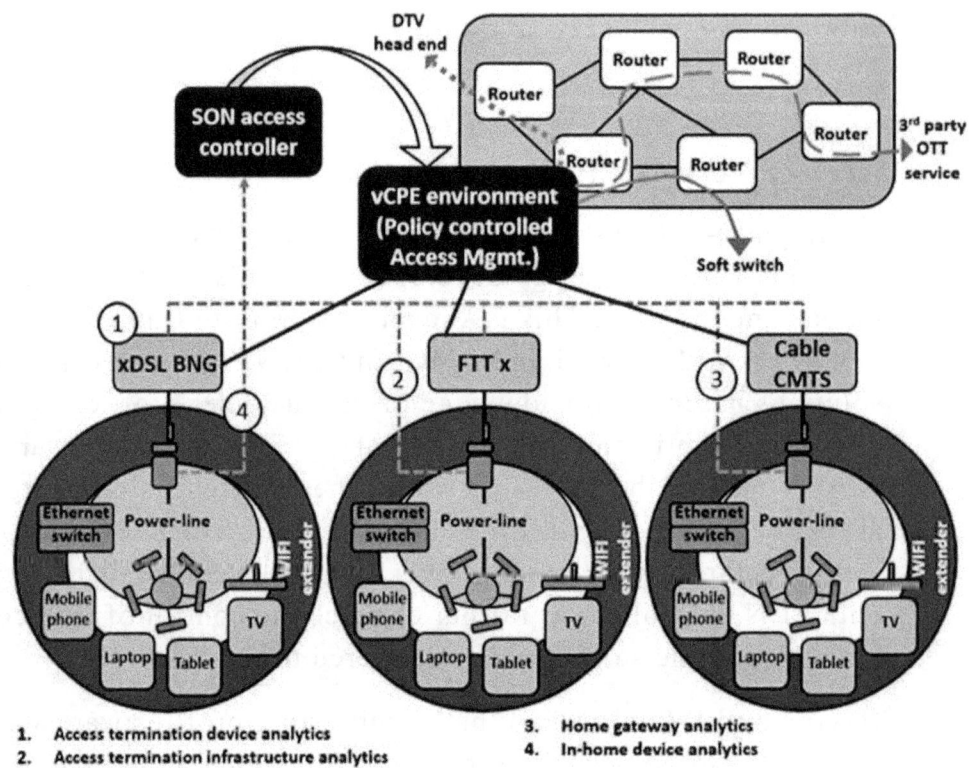

1. Access termination device analytics
2. Access termination infrastructure analytics
3. Home gateway analytics
4. In-home device analytics

Figure 10: Unified access technology control

This could lead to reduction in management and control access technologies and could drive a more widely accepted approach for future evolution. In the past, many companies worked with singular technologies as their mechanism to deliver to customers. This created terms, which we all use to refer at a high level to the technologies they were expected to use, e.g., mobile (GSM, 3G, LTE), cable (DOCSIS), PTT (DSL) etc. With current moves in the market to merge different access technology companies such as cable, mobile, PTT and FTTx companies, many who already have WI-FI infrastructures; it will be necessary for efficient product creation and service delivery to develop a unifying approach to access network control. By integrating these technologies through e.g. a vCPE solution, services could be delivered using an overlay construct. This approach could be permitted to interact with a localised modelling function for control of access network segments.

Virtualised CPE/In-home Management

Virtualisation of the CPE is the removal of some of the many functions from the physical CPE, and the shifting of these functional capabilities to a Cloud-based system higher in the network infrastructure. This allows some of the limitations of the CPE to be removed from the device, therefore permitting for the operator to turn up new functionalities without having to swap out the device. As a result, new service delivery is sped up, the costs of having to purchase a new CPE are reduced, there are savings on the logistics costs of getting a new device to the customer, and customer disruption is removed.

Through the use of COTS or the forwarding plane equipment (or a combination) greater functionality can be deployed. With the ability to enable greater processing and analytics capabilities on this system, the vCPE provides the opportunity to enhance the customer service delivery and management through access to greater analytics and the ability to enable new applications. By having a localised policy enforcement point, this can provide additional controls of all the access networks, irrelevant of access technology (mobile, FTTH, cable, DSL, Wi-Fi etc.). Not only does this enhance the customer experience, but also the capabilities of the network operator to better manage the customer experience. By having greater processing capability available in the local Cloud, many of the limitations of the customer device can be eliminated. In addition, this minimises the need for the consumer to have in-home security devices and routers. Today the majority of customers don't have the skills to maintain this equipment. In general, these devices are not maintained with the latest software releases to protect against intrusion by hackers, and constrained CPU and memory reduces the firewalling capability. By moving this functionality to a localised

Cloud, this limitation can be addressed by skilled staff who are up-to-date and aware of latest releases, bug fixes, and bring expert knowledge on the configuration rules. A line termination NTU is still required in the customer premises, and all traffic from a single customer can be encapsulated, and the layer 2 domain can be extended to the vCPE and PEP for the termination. By developing a full multi-access-technology SON architecture for access network management and control, the capacity could be fully managed to the end user device for their services. This permits for the virtualisation of functions such as the Wi-Fi controller software. The antenna will still remain on the CPE. The controller can now sit in the Cloud and can be used to permit full management of the channels utilisation. By reporting this data back to the PEP layer (vCPE), this information can be analysed and used to ensure delivery occurs on relevant channels where the signal quality and capacity is available.

Carrier Grade Equipment

Restructuring the importance of some protocols within the forwarding device enables the opportunity to create network elements that are carrier-grade. With carrier-grade network elements comes the ability to use less equipment per site, and the possibility to develop optical network equipment using protocols such as GMPLS or T-SDN. With having access to the C&C LNMF functionality and its end-to-end view on the network, this adds the intelligence to proactively identify problems and to take corrective action before they impact the infrastructure or services. Carrier grade equipment therefore becomes a building block in the end-to-end architecture. This becomes feasible when the other off-line functionalities are considered within the structure of the SDN Strategic Architecture.

In addition, virtualisation of the Customer Premises Equipment (vCPE) enables protocol and feature functionality to be removed from the PE router. When overlay SDN VPN is utilised, there is no further requirement for Layer 2 and Layer 3 MPLS VPNs to be configured into the PE router. Where SR is utilised instead of RSVP TE/LDP to control the flow of traffic through the network, this negates a considerable amount of functionality on networks. This creates the opportunity for carrier grade PE and P routers being achievable as older protocols are removed through migration.

Summary

This technology used in the SDN Strategic Architecture is not completely new as it utilises many long time existing technologies. Its aim is to simplify how services are delivered through focusing on management and control of the services, while optimising services based upon customer demand. The SDN Strategic Architecture aims to deliver a service aware networking capability by:

- Simplify IT/OSS through a centralised C&C function

- Enabling data analytics to drive network and service management and control

- Enabling more efficient use of the network, based on customer demand and network changes

- Automating customer service requests and utilisation through the C&C and vCPE/PEP functions

- Utilising virtualised and unified service platforms for cost reduction

- Simplify and virtualises CPE and STB components in the home to reduce cost, and simplify service assurance and network delivery

8.   SDN: Service & Infrastructure Architectures

By separating the control and the forwarding plane, SDN opens up the opportunity to create many new simplified and flexible approaches to operate network infrastructures. Many of these architectures are already in use around the world, and are being trialled by various network operators and enterprise customers. This chapter highlights the new capabilities and concepts that are and have become available, with the aim of discussing these concepts, rather than defining each one in working detail.

The control mechanisms introduced through the SDN Strategic Architecture create a structure that allows for the integration of management tools that support both the operations teams and the engineering teams with achieving their deliverables. Also being developed are many of the functional systems that evolve from these architectures and how they link to network, customer and service management systems.

SDN: Service and Infrastructure architectures

To discuss the forthcoming architectural changes, first a key point has to be considered. This concerns the concept of separation of control and forwarding. In existing networks, the infrastructure device has both functions in the same device, and the same vendor delivers this integrated capability. By separating these two functions, the programmability of the forwarding plane becomes open and permits the use of other protocols and offline applications to deliver and influence the creation of the FIB tables for the device.

This addresses two key architectural changes:

- The first is the programming of a FIB that has historically be performed through a closed API. As this is now open, other models for end device configuration can be considered.

- The secondary implication of this change concerns how data is exposed. Detailed analytics gathering can now be collated and used without being constrained to protocols such as SNMP, syslog, IPFIX etc. This is a primary focus of the development work being done on SDN but not all solutions being developed solve this critical issue.

As greater volumes of data now can be gathered, specific analytics can be produced by the network elements to create a detailed understanding as to how the service and protocols are operating. This can be done using APIs. As the API is bi-directional, this mechanism gives access to real-time data from the device and uses less processor intensive methods for gathering relevant data. As SDN now provides the ability to program the forwarding device using open APIs from external devices, this provides the capabilities to drive intelligent automation and to overcome the inherent limitations of unreliability of the IP Protocol. The analytics can be used as source data for the offline C&C systems. An example of this is that the LNMF calculations can now be used to define forwarding decisions for management of the end-to-end service. The traffic path can then be set for how traffic will be forwarded through the network using Segment Routing. Future manipulation of traffic flows can be delivered using information set in the NSH and IP Meta Data protocols. These markings will include the capability to include preferences relevant to the end customer service delivery selection.

To explain how these technologies drive new architectural choices, the following sections in this chapter discuss solution architectures options, that provide new control and forwarding techniques for networks when compared to what is available in today's networks. Most of these techniques only become feasible because of the centralised control and distributed solutions that are introduced through the SDN Strategic Architecture.

8.1   Underlay and Overlay

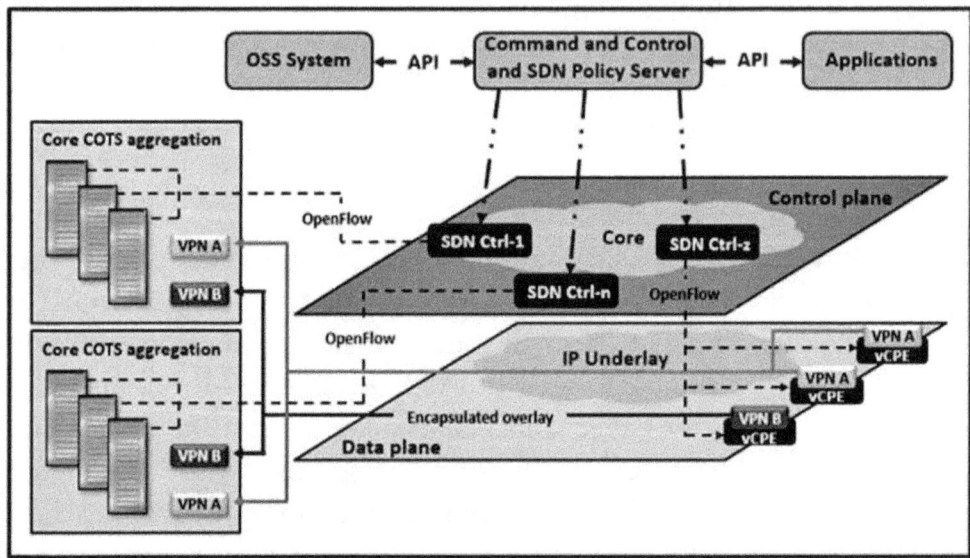

Figure 11: SDN VPN control plane, Overlay and Underlay

Underlay and Overlay are terms which are used to clarify the difference between the infrastructure, which is used to forward the packets, and an encapsulated mode, which is used to isolate the traffic for separated control through the network.

Underlay

The underlay network is a term used to identify the physical infrastructure of the network. The underlay (physical network) focuses on the moving of packets to the required destination. Some proponents of SDN see the future of the underlay to verge on being dumb and to only have the focus of forwarding packets as fast as possible.

This approach usually expects that capacities on the network are unconstrained. However, geographically dispersed core networks are constrained. With this in mind, the SDN Strategic Architecture identifies a solution that proposes control of both constrained underlay and overlay networks. It incorporates NFV functions through the NFV orchestration layer and enables new functionalities to address limitations at this layer. This ensures the ability of the overlay virtual layer to deliver the service with the flexibility it needs in order to meet the SLA.

Technologies such as WSCON or GMPLS can utilise dynamic optical capabilities to ensure convergence of control of not just the IP layer but also the optical networks. By using the Live Network Modelling Function, multi-layer analysis is able to identify situations on the network or service, thus permitting for intelligent decision-making. This gives the ability to trigger new optical or IP paths into the network through the policy or network controllers and permits for alternative paths to be identified and traffic to be forwarded according to the needs of the service level agreement.

Overlay

An overlay network flows over the infrastructure (underlay) network. Within the SDN Strategic Architecture an overlay network is the encapsulated instances of flows that are isolated, and which external controllers and intelligent tools select the control of their forwarding dependant on the business logic required for the service delivery. Currently in SDN at the time of writing (late 2014) these flows follow normal forwarding rules, but in the future these will be externally influenced using Network Service Header, IP Meta Data, Segment Routing and the LNMF. This will permit the forwarding path selection to be based upon the needs of the service and not to force it to only follow the global routing table enabled on the network. This evolving situation permits external sources such as the LNMF and network or policy controllers to identify particular flows and to enforce rules that are relevant to the SLA of the service.

This separation of functionality and the elimination of the need to hard-code the virtualised configuration of the service into the physical network, permits organisations to redesign and upgrade the physical network as needed, without affecting the overlay topology. This gives the organisation the flexibility to initialise services at speed, to modify services and to remove connectivity across the networks without having to interfere with the physical equipment that makes up the infrastructure.

NSH, IP Meta Data and SR, by providing extra information, allows for control of active flows. When these protocols are used in conjunction with the C&C systems, this enables the control of the overlay while interfacing to the constrained underlay network. This drives towards the goal of achieving real time managed service delivery.

8.2   Software Defined Data Centre Architecture

SDN Data Centre technology changes how Data Centres are operated and permits a new approach as to how the business can interface into the Data Centres when initiating new services. It also permits for multiple Data Centres to be created and controlled from a centralised control point. SDDC (Software Defined Data Centre) also enables an approach to service delivery that is very compatible with Agile and DevOps working methodologies, which match the desired business demands of many operators. To explain how the SDN approach differs, the following comparison is made between existing technologies and SDN technologies to identify the effort required when delivering comparative services.

Creating a Traditional DC service

Establishing new service connectivity usually involves a network architect defining an architecture based upon the customer specific requirements for the solution. The solution that will be developed is usually based on the knowledge, the experience and the personal technical preferences of the network architect. After this design has been handed over, an engineer investigates the current infrastructure build and then defines a design. The design may or may not follow the architectural rules and strategy, depending on workload and experience of the engineer. If standards are not followed, this will cause problems for the operational team, as they will be required to support a variety of solutions. This requires the engineer to identify and validate the command line associated with the vendor proprietary solution. To do this switch configurations are prepared, routers configurations are prepared, IP addresses are planned and allocated, change requests are raised, and the network engineer brings all of this together. This stage of the work is completed according to the experience and knowledge of the engineer with the proprietary systems upon which they are creating the solution for.

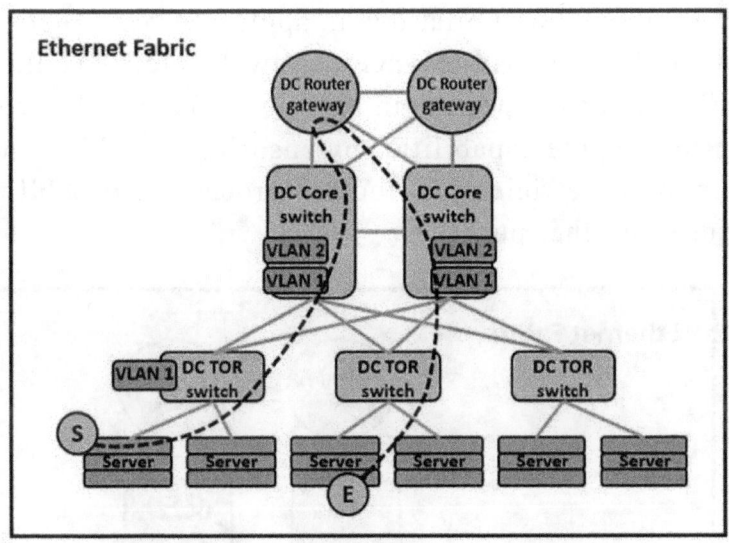

Figure 12: Traditional Data Centre service configuration

When this has been completed, validated and tested against the latest proprietary releases of software, change requests are raised, reviewed and eventually approved. The engineers then implement the solution, sometimes in the middle of the night when a person's brain is least awake or aware.

The whole process may take weeks and usually starts all over again when the next project comes along. The quality and completeness of the solution that is delivered is dependent on who is assigned with the task, their workload, their preferences and their skill levels. In addition, the solution defined has a technology focus based upon the capabilities of a vendor proprietary technology, rather than a focus on the needs of the business and the business solution that is required.

SDDC Architectures

Software Defined Data Centres were the first focus of SDN. Focus on Data Centres existed because they are less constrained environments, and because of the businesses need for automation, which had been enhanced with the significant development of Cloud functionalities. These new requirements highlighted the limitations in the current networking methodology. When the benefits of the evolving SDN technology were realised, this drove a focus on SDN development in the Data Centres. As SDN development progressed, it highlighted the limitations in appliance-based network functional systems such as load balancer, firewalls etc. This drove the significant developments in NFV. NFV helps release stranded Data Centre compute capabilities by ensuring that this costly and sometimes inefficiently allocated processing capability is available for use for other purposes.

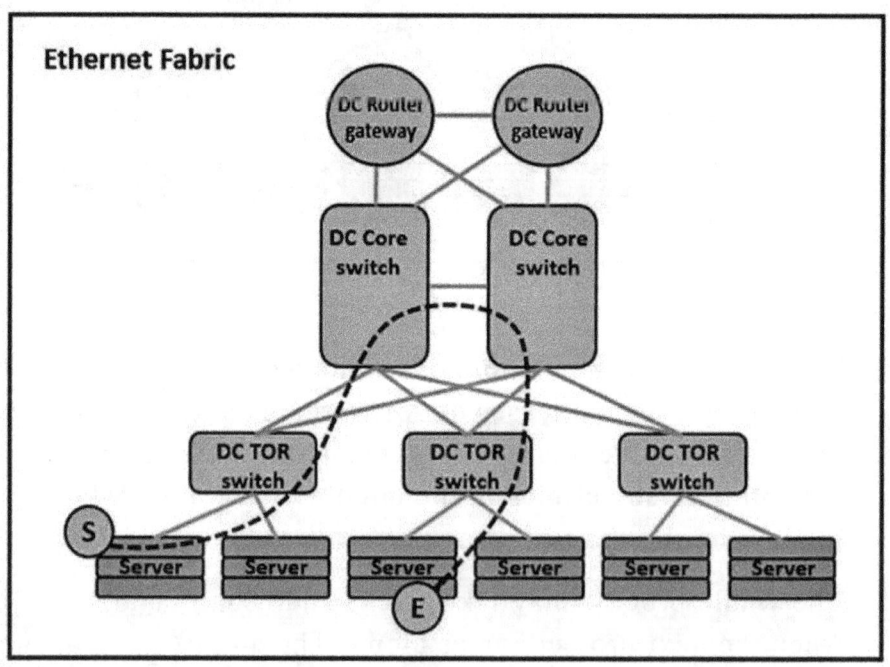

Figure 13: SDN Data Center overlay principles

By having the capability to trigger new instances and to bring these automatically into play, new architectural structures can be considered for redundancy and location. As these can be spun up on COTS hardware and use overlay technology, complex network configurations are no longer required to address failures, as a new structure can be instantiated to overcome a failing solution.

SDDC releases the power of virtualised computing by enabling the network connectivity through simplification, automation and by providing the extra focus on network management. This involves the breaking down the restrictive network boundaries and zones that have made creating network connectivity for the compute capability difficult, manual and time consuming. Additionally, multi-tenancy capabilities within SDN ensures that the different Cloud clients with varying service and security requirements are capable of being enabled with the appropriate configuration as required.

SDDC uses overlay instead of underlay. Most SDDC solutions don't configure service-specific parameters into the underlay network elements. Using SDDC policy control automated configuration techniques ensures that standard services can be network-enabled through an SDDC service portal. Via this portal a customer simply chooses the options required for their services. Then the configuration for these services is applied, and the services are enabled within a matter of seconds or minutes depending on the solution required and the technology build in place. These can be enabled based on time-based requirements, or simply enabled and disabled at the customers' time of choosing.

With SDDC multi-tenant capability, the consumer of the resources can control and utilise shared infrastructure with the ability to use services based upon Cloud computing and storage. In a SDDC environment, separation and isolation of the network is provided through overlay network capabilities using technologies such as VXLAN, MPLSoGRE, NVGRE, etc. The technology used is based upon the direction the SDDC vendor chooses or the direction that the operator requires.

Additionally, many SDDC vendors provide virtualised router functionality that allows for breakout capabilities to the core and controlled layer 3 connectivity between instances. This router virtualisation operates on a virtualised instance. Logical switches exist on the VM's and provide for establishment of the connectivity using overlay tunnels between the instances and the network functions, such as the firewalls or load balancing.

Creating an SDDC Service

When working with a SDDC solution, a customer can activate the solution they need using a service portal. The network engineer is required to have created the templates/configuration snippets based upon the service requirements being provided to the customer. Some SDDC solutions use policy based controller template based configuration solutions that are created for each requirement of the solution. These templates are structured so that they can be reused and are created through GUIs. This reuse permits for speeding up solution development and reduces the needs for testing, as many aspects of the templates used will already be live in the network. This aids security within organisations and simplifies the delivery of security controls for the solutions that the organisation requires.

When a hypervisor application is instantiated, the corresponding network configuration is derived from the engineers' pre-prepared templates and the network is automatically set up according to policies defined. This permits for automatic configuration, as the configuration is only applied at the end points of the Data Centre network.

As overlay networking is used in SDDC, none of the existing intervening devices are required to be configured. End point configuration is enabled using virtualised instances of the end points when the application is triggered using the Cloud orchestration system. In reaction to the application being enabled, the policy controller applies the appropriate policy based templates for the services into these end points for network connectivity. Using the underlay routing enables connectivity, which has been in place since the infrastructure was installed. VXLAN and other protocol tunnels can be created between the end hypervisors, thus permitting the connectivity to be established either within a Data Centre or between geographically disperse Data Centres.

Within the template all features and functionalities can be defined. Examples of these are the forwarding rules, the security entries, the port configurations, or the VXLAN identifiers. For rules that are variable and per site dependent, e.g. access lists, VXLAN identifiers or IP addresses, and these can be allocated by the resource management application using an API. Business and configuration logic within the resource management application can be used to set access list rules for the IP addresses.

With some vendor developments, the network controller then uses OpenFlow to signal a forwarding table to the OVS stack running on the hypervisor, thus enabling a forwarding rule set on the end device. Only signalling control traffic is forwarded to the end network controller, which derives a connectivity map for the hypervisor that is part of the same tunnel/encapsulation.

Once the templates have been created by the network engineer, the operation team can apply the templates on a per-service basis or the policy controller can be connected to a service catalogue portal, thus allowing customers to order and automatically instantiate a service within the Data Centre (including NFV and Cloud resources). This process can reduce the rollout time by weeks. Depending on the service catalogue and the variables it defines, the same template can be instantiated in a variety of configurations for different customers using different resources. The multi-tenancy capabilities that exist within SDDC solutions ensure that each customer instance is independent, unique and in line with customer requirements.

When a new service is required, the policy controller validates the network resources against the resource management application and automatically allocates appropriate resources. The SDN orchestration ensures that NFV and Cloud orchestration systems allocate and remove the resources dependent on whether the order is to disconnect or allocate a service.

8.3   Software-defined VPN Architecture

In the world of overlay and underlay, over-the-top VPNs-based services become a natural solution. They don't require configuration into the network infrastructure as currently required with MPLS VPNs. MPLS layer 2 and layer 3 VPNs which can be difficult for the network operator to maintain and require a compliant standards-based solution that brings complexity especially within a multi-vendor environment. SDN VPN offers the following capabilities to the operator and consumer:

- Similar in creation as to the historic VPN technology using IPSEC/GRE etc.

- SDN VPN does not bottle neck data traffic at a single point in the network and unlike GRE/IPSEC routes all data traffic directly from the data source to the data destination.

- Management statistics are gathered from the end-points for significantly greater management functionality.

- SDN VPN only forwards the light weight signalling and management traffic to the centralised network controller.

- Rapid service deployment without constraints by the network infrastructure.

- End-to-end provisioning independent of the underlying network or transport

- End-to-end application and traffic visibility and control

- True multi-tenancy for layer 2, layer 3 or even layer 4 or per LAN port services

- Resource management tools automatically manage the allocation of resources

- Eliminates scaling issues with VLANs numbering

- Increased flexibility to add new applications using hosting capabilities, e.g. PABX, remote desktop, share drives, etc.

- CPE plug and play

- Detailed service analysis and reporting

- Flexibility with development, management and control

- Simple integration into existing MPLS VPN solutions

- Limited requirements for appliance based CPE: CPE based on e.g. X86 COTS hardware

- vCPE can be enabled either onto a COTS capable network element or server

- Off-net site connectivity is established using IP connectivity

- Removes configuration mistakes through use of policy controlled template based configuration

- Where connectivity exists, services can be turned up in minutes across multiple sites through use of virtualised instances

- Existing services can be modified in minutes through applying a new template

- Customer managed LAN services becomes achievable

The SDN VPN approach changes the historic rules expected by network engineers for building VPNs and redefines the capabilities of what can be offered within the business VPN market.

Creating a traditional VPN service

The traditional development cycle for VPN usually involves a network architect defining an architecture based upon the customer specific requirements for the solution.

After this has been handed over, an engineer defines the design and identifies the command line associated with a vendor proprietary solution. To do this, switch configurations are prepared, routers configurations are prepared, IP addresses, VRF's, VPN-ID are planned and allocated, change requests are raised, and the network engineer completes these many stages of work.

Next, these need to be inputted manually to the customer provisioning or management systems. When this has been completed, validated and tested against the latest proprietary release of software change, requests are raised and approved. Individual CPEs are identified, RFPs are run, equipment is tested and validated, etc. When the right CPE is identified, an order is placed (if the equipment is not held in stock), and this is shipped to the customer's premises. During this stage, there is usually considerable backwards and forwards communication about which CPE is the most suitable with regards to the constraints of the various CPE types and the cost effectiveness of choosing one over another. The likelihood of up-sell after the customer is connected is also taken into account. Then the implementation engineers implement the solution. In general, the whole process takes weeks or months. In addition a feature check is required against the current live network software version and where not compliant the all network elements would require upgrading to support the service.

When a new customer with new requirements is identified, the processes starts all over again or, in companies which use a building blocks approach, a few weeks are spent validating old designs against new customers' requirements to check if the existing solution fits against new models of CPE.

Creating an SDN VPN Service

Depending on the vendor solution chosen, the SDN controller and SDN applications that are used for SDN VPN can be the same that are used to manage and run a SDDC solution.

For SDN VPN solutions, a network designer defines a template for the VPN product through a GUI. Templates are reusable and can be defined for the usual combinations of VPN types. These include - but are not limited to - the usual three types; Layer 2 P2P (Point to Point (EoMPLS)), Layer 2 P2MP (Point to Multi-Point (VPLS)) or the common layer 3 VPN. In addition, SDN VPN solutions permit application-based layer 4 VPNs on a per-LAN interface or per IP-port VPNs to be defined. These VPN types can be configured based on templates created for forwarding rules. Templates that are based upon forwarding rules enable a greater level of security within organisations and simplify the delivery of the security controls. Templates are required to be pre-prepared by the engineer.

When an application is instantiated, the corresponding network configuration is derived from the rules defined within the template and the network is automatically configured according to policies defined by the network controller.

This permits for automatic SDN VPN configuration with no network touch as the configuration is only applied at the end points of the network and none of the intervening devices are required to be configured. This approach removes the need to validate existing router software for bugs particular to the service features enabled, to change out router hardware because of limitations in that equipment, or to change the configuration on the PE router.

The networking end point creation is done using of the customer device; the configuration is applied to these end devices. Using the routing that exists over IP layer enables connectivity. Encapsulation tunnels are created, and the end device routing ensures that the traffic leaving the CPE always utilises the shortest path permitted by the GRT to be followed between VPN sites. Virtualisation can take place on COTS hardware at the PE edge aggregation layer of the network or on COTS based CPE.

This provides the additional capability to load services on the COTS hardware, therefore avoiding the need to end new CPE to the customer premises when new features are required. In the case of COTS e.g. X86 based CPE or servers, relevant applications can now be spun up on the CPE if required.

Within the template all features and functionalities are defined. Examples of these are the forwarding rules, the security rules, the port configurations, or the VPN identifiers. For configuration rules that are variable or site specific, e.g. access lists or IP addresses, these can be either gathered through an API from an existing IP Address Management system or fully handled within the SDN VPN resource management system. Access list rules are automatically derived from the business and configuration logic in the template. When the end device boots and securely identifies itself, the controller pushes the configuration to the customer premises device.

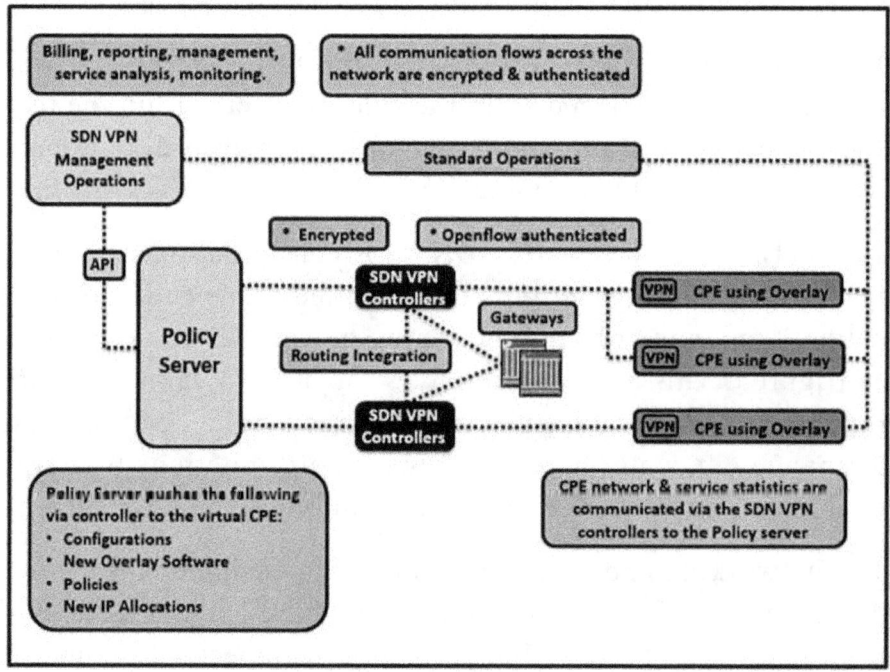

Figure 14: SDN VPN architecture

Only signalling control and management traffic are forwarded to the SDN Network Controller and its management applications. Amongst other things these management and control applications derive a connectivity map for the CPE to other devices within the VPN and this is used to create the forwarding rules for VPN connectivity and device connectivity. These forwarding rules are then programmed to the end device.

Once a template has been created for a service, it can be set into operational processes to directly trigger the configuration to the end device for the service, or it can be interfaced to an appropriate service catalogue portal. Dependent on the variety of options, it is possible to permit for the direct ordering of a variety VPN features with a variable number of sites by customers using the portal. For

94

multiple instances of the product, the same template is instantiated using different resources as many times as required within a multi-tenant environment.

The network advantages brought by SDN VPNs are not just the removal of the need to provision MPLS VPNs (and those consequences) into the network. Also, the need to have heavy s/w code placed on routers for MPLS VPNs becomes obsolete when using the overlay SDN VPN model. It also removes the necessity for routers to internally manage the MPLS VPN etc. resources, or the need to validate and upgrade software loads to fix bugs or security problems introduced by MPLS VPNs (be they layer 2 or layer 3). Reducing the complexity of routers helps drive forwards the concept of carrier grade routers and brings with it: A reduction in electricity consumption, increased throughput, increased integration to the optical layer, reduction in cost, and greater capabilities for future network equipment evolution. This permits moving from the methodology of focussing on infrastructure management as a method of emulating the service management to focusing on the overlay and ensuring the management of the service carried within that overlay.

The benefits from a business point of view are highly significant: Through a full SDN network and policy controller, services (on hypervisors or CPE) can be enabled and activated in minutes rather than weeks. For CPE this is dependent on the availability of fixed line or wireless connectivity to the customer premises.

SDN VPN permits the operator to enable the customer to take greater control of their infrastructure. Via a service portal, the customer can be permitted to reconfigure their infrastructure according to the template provided. The service portal also permits the customer to order new capabilities immediately. This gives the customer considerable choice and flexibility to be able to immediately modify and change their products at the time of their choice permitting them to move in tune with the needs of their business and not the existing network operator infrastructure change request processes.

Safe guards built into the SDN VPN systems and service catalogues minimise the changes the customer can make. Steps can be built into the service catalogue or operational work processes to validate and to check the business logic and technology logic of the requested changes.

8.4   Virtualising the CPE

The function of virtualisation of the Customer Premises Equipment (vCPE) sits largely in the environment of the work that is being done by the NFV. However, what it produces and how this can be utilised within an SDN controller environment creates new product opportunities and capabilities for the network operator, the customer, customer security, service delivery and service management on the network and the Internet. The concept of vCPE suits both residential and business customers' needs and fits well with automation through the OSS/BSS using a service/product catalogue.

With virtualisation of the CPE, the focus is on the removal of some of the service functions that exist in current CPE. Functionality is removed from the device and shifted into the Cloud. The Cloud environment will be deployed at a relevant position in the network depending on the choices of the architects. The Cloud does not have to be a single Data Centre or even a redundant Data Centre – it can simply be a few Cloud orchestrated COTS devices residing beside each PE router at dozens (if not hundreds, depending on size of the network operator's infrastructure) of points in the network.

By removing this functionality from the fixed appliance based CPE, the operator is no longer constrained by the limitations of that CPE and the amount of memory or features that the CPE can support. This provides a very significant flexibility when enabling new features and services.

The virtualisation of the CPE is made feasible using tried and tested technology, and is enabled through functionality that has been around for many years. This includes layer 2 encapsulation methodologies, network virtualisation functionalities, Cloud functionalities and additional NFV capabilities such as the virtualisation of firewalls.

Customer Benefits derived from vCPE

Moving the function from the CPE enhances the service capabilities for the consumer, especially when used with a service catalogue portal. This permits the consumer to gain accurate management capabilities and to make use of the provided flexibility to modify their VPN setup or their residential service.

Using a vCPE capability, the residential customer no longer is required to purchase in-home routers, power up additional in-home devices such as set-top DTV boxes on a 24 hour basis, manage their own security, set their own firewalls rules, attempt to debug issues they are experiencing or manage the upgrade of software loads for their own CPE. These and many other complex technical functions can now be handled in the Cloud by the operator. This removes the need for the residential or business consumer to personally control a significant volume of the complexity of the service. This is especially relevant as the majority of customers are not technically qualified to understand the required configuration. This increases the level of security for the consumer; as many hackers access home user information through hacking the low end functionalities on home devices, as these systems generally do not have the advanced capability to handle complex attacks.

For the business consumer, the vCPE provides the ability to have the network align with their business needs at short notice, therefore improving their opportunities in the market and enhancing the operational efficiency of their business.

Operator Benefits of vCPE

The vCPE function permits for services to be created and run in the Cloud. This brings many benefits for the ISP: the opportunity to sell inexpensive Cloud based applications and services that are directly connected into the business or home. This offers the ISP the possibility to extend the lifetime of the physical CPE in the business or home, as removal of functionality from the device allows for freeing up of CPU load on existing equipment.

Enabling the considerably greater scale of computational capabilities available in a Cloud environment improves customer management. With the ability to do greater analytics through APIs, the ISP can resolve network fault proactively, which will greatly improve the service the consumer receives. With these capabilities comes the possibility to enable personal services in the Cloud, which greatly improves the ISP's opportunity to create new revenue opportunities, while at the same time allowing for test environments for new products to be quickly trialled. This enables a new wave of innovation for vendors and operators to address customer management.

Migration to a vCPE solution can be done granularly. Depending on the ISP's requirements, they will be able to initiate a migration to a vCPE solution based on a very varied number of technical choices. This gives the flexibility to move forward without having to invest heavily on OPEX and CAPEX. This will require working with the install base of CPE vendors, and to have them provide a replacement code for their CPE, that enables an encapsulation of the layer 2 traffic from the vCPE to the Cloud environment.

For this to fully operate, full redundancy and failover capabilities are necessary, and the COTS solutions should be scaled to the required number of customer instances. The setup and build requires interfaces from the service catalogue into the SDN policy/network based control solution. Using the APIs enabled in the SDN solution permits for services to be provisioned and modifications made to customer preferences, which can be initiated via a customer portal. These changes can then propagated down through the mediation layer and the policy and network controllers, where the changes can be invoked within seconds of the service being enabled by the consumer.

Such changes will provide the ability to rapidly enable new service deployment without requiring changes to the network infrastructure.

Service management benefits of vCPE

It should be noted that virtualisation does not only mean the positioning of all functionality into the Cloud; it can also be enabled on a single COTS based network CPE device or customer premises server. Such devices can be used to include multi-instances of functionality through programmability to initiate various different services out of one device. This way, both residential and business customer can initiate and consume services in a highly flexible manner.

In such solution an SDN policy and network controller can be used to program the end point for enterprise services using open source device programmability such as OVSDB. Alternative solutions are available and OVSDB is listed here as an example. This open source software is capable of setting the required forwarding instruction set for the hardware on many COTS hardware. This permits for a standardised approach to network management, while permitting for an open and flexible hardware sourcing from a variety of hardware vendors and processor manufacturers.

When considered in isolation, vCPE capabilities, along with SDN capabilities, provide a significant improvement in service delivery for all parties. However, when considered not as an end goal but as a starting point, the capabilities to provide future functionalities and capabilities become very significant.

Evolving access network customer service control

SON technologies are under development within the 3GPP group with a focus on RAN networks control and management. The principles of this solution are also being considered for WIFI networks. For the access network, SON technologies could be further developed to allow for the control of the overlay network by permitting isolating the control of the access network from the physical access technology. By gathering the underlay infrastructure analytics and evolving networks using SON capabilities these can control and manage the underlay and inform the overlay on everything from load, errors, quality of link etc. This provides the ability of the network controller through integration with the analytics reporting to set the path of the service onto the appropriate WIFI channel in the home, to select the appropriate DOCSIS channel, to select the most appropriate WIFI hotspot, to steer the packet through a RAN network, etc.

With the ability to position the traffic onto a selected path, comes the ability to select which technology the traffic should take. In the future, products could be defined that set the path the traffic will follow depending on the service the customer has purchased.

Traffic could then be off-loaded efficiently during moments of peak load, depending on the priority of the service and the availability of capacity on one access network or another. With NSH/IP Meta Data comes the ability to fully enable the vCPE controller and to have the ability to proactively steer the flow over the appropriate access network. As the network controller and its applications will have knowledge on the destination, the requirements of the stream and the business logic of how the user will be treated depending on the product they have purchased, greater service control will be feasible. The developing and evolving of management and control capabilities of access network control, fits with the changing business models caused by the needs to supply Quad Play and the change by operators who now operate multiple access technologies to deliver services to their customer base.

Core network Customer Service Control

With a vCPE environment the network operator achieves greater flexibility in reacting to the consumer's choice. Where a consumer has requested a service to be delivered with quality, (having marked this option via their customer portal) traffic can now be steered onto a Segment Routed (SR) path. Segment Routed paths are not

100

initialised on a per-user basis. They are pre-allocated and pre-modelled paths that have been set up through the network to deliver specific services according to the specific requirements of the service using the unreliable nature of the IP Protocol. They aggregate traffic from customers who have similar requirements and needs. These are then managed across the network to ensure that the paths remain unconstrained according to the needs of these aggregated services and therefore limit the unreliable nature of the IP Protocol. Management modelling control visibility is provided by the LNMF. With its ability to analyse current data on load and having access to trend and historic data, the LNMF will be able to accurately map the SR path into the appropriate path on the network. Using analytics gathered about the network, the LNMF will be aware of the characteristics of each forwarding device, link etc. and its capability characteristics. This imposes external end-to-end service awareness on the flow from the network Command and Control system.

Using parameters exposed through the OAM function in the Network Service Header (NSH) to the network management function and network modelling function (LNMF) the latency, packet loss, delay and jitter will be exposed. Should these fall outside a predefined threshold the LNMF will advise on a separate path and the network controller could inject a change of path to the relevant segment routed tunnel source router within the network. This provides the control of the on-net issues for the core of the network.

8.5   Virtualising the Access Control function

For the customer the primary consideration is the service, not the connectivity technology.

The vCPE system and SDN controller ensure the services can be delivered across any access technology; hence these systems permit the SDN policy and network controllers to take over the full running of all access technologies.

By virtualising the base functionality of these access technology controllers into the COTS hardware (stripped down technology connectivity functionality devices are still required to physically connect wire-line and wireless network infrastructures), this permits the access controller functions to be run as part of the control of the vCPE solution. This reduces the need for all access networks to be treated differently, and unifies the control and management of multiple access technologies within a network operator's infrastructure.

By processing the analytics gathered and by utilising SON-type functionalities, traffic could be steered to utilise whatever relevant access network is available to connect to the consumer. This integrates the control of regionalised RAN or WIFI networks into fixed line networks, where the service can then be delivered through controllers using core technologies such as SR to provide the latency and time sensitive connectivity to the other regional vCPE domains.

8.6   Dynamic Optical

Optical networks have long been viewed by IP Engineers as only being relevant to IP networks in as such as they provide the underlying geographical logical or physical connection between the routers. Separate teams who work with separate vendors products have historically managed optical networks. With optical networks, complexity exists up to the edge of the laws of physics, and as a technology it must operate in the field and deliver error free connectivity. It is highly complex to create, develop, design, implement and to operate.

Optical Networking has a lot in common with SDN. The SDN approach of centralising the control of the infrastructure and services has been the standard approach used for years in optical networking. Optical networks use an EMS to centrally control the infrastructure but have a no understanding of what (IP flow) services utilise the optical network. Significant change has happened over the last few years, and new thinking and more competitiveness are now driving a very new approach to delivering the optical capability and its ability to integrate into a multi-layer network.

GMPLS and other protocols have been developed to interconnect the IP plane to the optical plane. The goal of these solutions is to enable the ability to trigger new capacities into the network. This is limited in that the external management and control does not have visibility of the end-to-end solution but only of the connection.

The SDN Strategic Architecture, through incorporating the LNMF, gives the end-to-end service visibility by modelling not just the IP layer, but also the routers, the optical circuits, and the optical nodes on the network through to the ducts that carry the fibre.

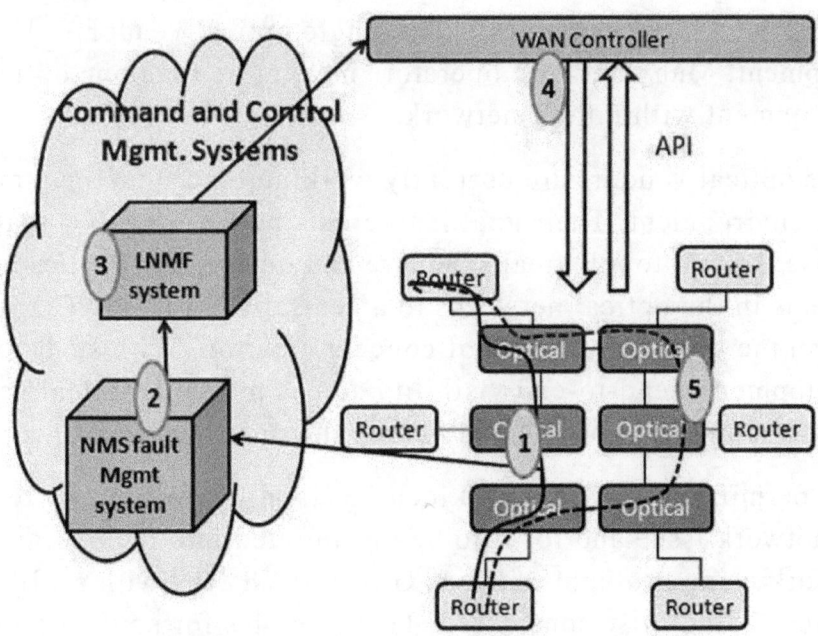

1. Error threshold exceeded, notification sent to Fault mgmt.
2. Solution request triggered to modelling function
3. Solution modelled and new path parameters to Controller
4. Controller programs new path into the network
5. New path is activated across the optical to appropriate routers

Figure 15: Proactive management controls using dynamic optical

Through incorporating this visibility of the network and through the modelling of failure scenarios, the LNMF can identify the appropriate solution that could be automatically triggered using protocols such as GMPLS or WSCON into the network.

This provides resolution to the problems faced by the network operator using information about the services, and not just connectivity. The issues with fault resolution using only a connectivity map is that the new circuit may significantly increase latency and can cause other issues as they do not consider the needs of the service. Data about the optical infrastructure and its resources can be achieved through interconnecting to the EMS and including having a separate LNMF capability in that node.

Although this approach provides a partial solution, it still maintains a proprietary environment, and for many operators it doesn't give visibility of the services running on their network, but only visibility of services that run on a single optical vendor's equipment. Many network operators have more than one optical environment within their network.

Some optical vendors are currently working on producing a more open environment. Their goal is to create new protocol(s) that expose the full topological structure and device capabilities of the systems in the optical networks to a centralised LNMF. This work is still in the very early stages of concept creation. With such developments, end-to-end visibility across multiple optical vendors' infrastructures can be achieved using the operators LNMF.

This permits for the LNMF to identify a valid new path, to resolve the network issue and for it to be instantiated into the operators' network using protocols such as GMPLS/WSCON with all IP level services being fully considered. This provides for new possibilities to become feasible with automation of management and control and for the deployment of optical capacity into the network. Work is ongoing in within the ONF on T-SDN (Transport-SDN). This has a lot of similarities as to what the author describes as dynamic optical.

The SDN Strategic Architecture approach permits the SDN management systems to monitor the network and to trigger demands to the infrastructure. An SDN Controller gives the opportunity to integrate with proprietary solutions as a first step. Opening up the APIs to the optical nodes and not just the EMS provides a mechanism of evolving the optical infrastructure to deliver more flexible capabilities within their technology to the network operators.

8.7   Self-Organising Access Networks (SON)

SON is included because it is seen as an interesting and enlightening approach that can deliver valuable business benefits if evolved into the SDN Strategic Architecture. This ongoing work offers significant potential when considered in relation to possible enhancements that SDN can bring. SON has historically only been considered in the context of radio access networks, whereas the SDN Strategic Architecture looks to highlight this work and to suggest the evolving of the principles of SON into fixed access and core networks.

SDN control and management solutions show a strong similarity in thinking to SON. Some vendors' SDN solutions have incorporated some of this thinking in the shape of building blocks for their SDN solutions. Additional work is needed, but with these principles in place this work will continue to evolve over time to drive towards the creation of a technology that permits for a self-managing, self-maintaining and self-expanding network.

To clear up the meaning of the term: The official meaning of SON is Self-Organising Network, however many in the industry use the term Self-Optimising Network. According to the 3GPP standards, the 3 primary functions of SON are self-configuration, self-optimisation and self-healing. Therefore optimisation, although a key function of the work being addressed, is but a functional subgroup of Self-Organising Networks.

Mobile operations and 3GPP are advanced with their development of SON due to the requirements of managing and controlling RAN environments. Although the focus has been on the RAN environment the same principles of control and management through self-configuration, self-optimisation and self-healing are as relevant for mobile as they are for fixed.

These aims are not revolutionary as they have been the goal of network engineers for decades and extend back into protocols such as IPX™ or AppleTalk™ where considerable success was achieved. These are also the goals of SDN, but SDN considers these not just for network infrastructure control and management but also for end-to-end service control and management.

Mutual Goals of SON and SDN

Self-Organising Network (SON) type capabilities have been the goal of many network operators and vendors for many years. Many approaches have been tried, even prior to the Internet in other protocols with relative success. However IP with its limitations has so far not achieved this potential. Changes being enabled in Software Defined Networking (SDN), Cloud, NSH and Network Function Virtualisation (NFV) now offer an alternative approach through the centralised controller model.

Although SON is not generating the same attention as the SDN/NFV developments, capabilities being incorporated into the SDN toolbox through protocols such as OVSDB, NetConf/YANG, the introduction of the modelling functions, the introduction of open APIs and many other mechanisms can help realise better control of the most costly part of the network, the access network. Through having similar goals these can now be integrated by design into the principles of SDN Strategic Architecture. This will provide for a new level or operational control over the network and will permit an end-to-end architectural transformation to take place.

All networks, no matter if it is an PTT's, an OTT company, a mobile, an enterprise etc. have experienced the same limitations for many years when it comes to:

- Programming the network

- Optimising the infrastructure for more effect forwarding of traffic

- Minimising the cost of the network

- Proactively enabling fault resolution

- Proactively finding faults

- Ensuring a service is fully managed

- Ensuring the service is delivered according to SLA

- Ensuring the service is configured correctly

- Ensuring the security of the service

- Turning up services when needed

Advanced SDN developments are employing some techniques that are aligned with the work SON creates for RAN environments. Such advanced SDN solutions also include new network functions such as real-time network modelling tools and plan to use new protocols such as SR, NSH and IP Meta Data to both enhance the information extracted from IP flows and to deliver enhanced IP management and control capabilities. In addition, SDN has the ability to deliver automated programmability of both the network elements and the end CPE using open APIs. When unified, these capabilities allow for proactive management and control to be applied into the access network.

Through these APIs, a SON approach could bring significant gains in operational change and performance of the underlay networks. Another major benefit is that these improvements will allow for the operator to concentrate more on the service rather than on the infrastructure management. When considered with NFV and the COTS virtualisation capability that NFV can use, this allows for new applications to be added into the architecture and scaled to addressing problems, which all network operators have struggled with for years.

Network management systems for the most have focused on collating alarms and events generated by the infrastructure and have propagated visibility of the issue. Unfortunately, these solutions have been hampered by the limited visibility of the technologies they tried to control, manage and report on.

SDN can benefit from the work already carried out in the development of technology developed with the goals of SON as their guiding principles. This discussion does not attempt to cover all topics and considerations relating to SON, as this topic would require several books just by itself.

When considering the goals of SON within the context of SDN, the 3 key functions and enablers are:

Self-Configuration

The self-configuration function gives the ability to activate a relevant instruction set via a programmable APIs using SDN network and policy controller capabilities onto a specific relevant network element. This allows for a network element to be programmed through a policy controller with a set of instructions, relevant to the needs of the service. This ensures the node is automatically reprogrammed when required during operations according to a standardised set of rules. This ensures that the correct configuration is always applied.

Network engineers are required to prepare and test the initial template configurations for the element and its relevant protocols. The automation of network elements can be triggered by a variety of sources. These include network engineers, operations staff, advanced network and service management platforms or a service or product catalogue from the OSS/BSS solution.

Triggering can happen from a management application within the SDN policy and network controller or from the portals. The choice of how this gets applied is based upon the architectural and business decisions made for the solution by the architects. This ensures the accuracy of the applied configuration, as only tried and tested configuration should be included into automated processes. As a result, the time to create change requests as a standard approach is reduced significantly and can be used for e.g. the installation work load.

Self-Optimisation

In SDN, the self-optimisation function doesn't just focus on the infrastructure, but on the service and its requirements itself. The self-optimisation function can receive instructions from the LNMF, and, using developing standards such as SR, will ensure the optimisation of prioritised traffic flows through the network elements on the end-to-end path. Instructions can also be delivered to redirect traffic to other nodes. This function is increasingly expanded upon, with consideration given to regionalise monitoring being fed into a regionalised controller within the vCPE environment. This regionalised control will ensure that the traffic directed across the local infrastructure addresses the needs of the service as it is handed off from the core, while at the same time taking into consideration the quality of the access networks. More parameters are required to be added to this functionality; this should include the ability to control the flow of data across multiple different access technologies, thus permitting for delivery of traffic over the service optimal path, dependent on customer requirements and in line with the needs of the service. This fits well with the changing nature of the industry, as the classic network operator model continues to go through change: Previously, network operators focused on single access technologies with cable operators using DOCSIS, PTT's using xDSL, mobile using xG technologies, FTTx being the focus of start-up operators and PTTs, etc. Now the industry is crossing over, because customers require quad-play services. Additionally, companies now are investing in or purchasing companies that have historically been considered outside their area of focus. To operate such a multitude of access technologies without having to run a multitude of different technology departments requires a new access technology management approach - that is discussed in the SDN Strategic

Architecture. The alternative is an even slower time to market and an increase OPEX costs. Competitiveness will be damaged, because product innovation and launching will only be possible when the entire footprint is aligned with all features and capabilities. If this is not done, it will take years to bring new products to customers, in particular in larger networks.

Self-Managing

The concepts of a self-healing network from SON is included within the architectural thinking of SDN thinking, however SDN thinking also incorporates a self managing and the ability to self-expand. These functionalities all use the same analytics gathered about the state of the infrastructure and service. Using the information gathered about the needs of the service through SR, NSH and IP Meta etc., events can be triggered to respond to the issues within the network across the multi-layer infrastructure. When service degradation is identified, e.g. through the NSH OAM measurements, an optical path fails etc., the modelling function can be used to calculate an optimal route to reposition the service on the network infrastructure, thus automatically overcoming any possible degradation. This will reduce the severity and duration of outages, and lead to measurable deliverable SLA's and higher customer satisfaction.

A scenario

If, for example, the user had requested a video stream that was 12Mb in size and lasted for 80 minutes, this relevant service information could be marked at source in one of the context fields of NSH/IP Meta Data header. When receiving this information, the vCPE solution could check the access network analytics and its user database. If volume-based billing was enabled on an access network, and if the consumer had requested that they never wanted a stream of greater than 4Mb, then the following actions could happen: A transcoding application could be used to reduce the stream size, or a signal could be sent to the consumer warning them that they had initiated an oversize stream. Following that, the user themselves could request the adjustment of the stream size from their content provider. The SDN Network Controller would forward the service onto the appropriate path over the appropriate access network, after it had checked against the SON function to ensure that the capacity was placed on an appropriate access technology, both on the in-home network and across the access network. In addition, it should be possible to create a business model with the content provider to eliminate the content the customer consumes from their volume based billing model.

8.8   Interworking across the Internet

The use of a C&C system provides the opportunity to use business logic for the enhancement of service control using Quality of Experience on traffic between different ISPs. This can permit users to request a quality-based stream that traverses multiple provider networks.

The majority of the traffic on the Internet today only needs Best Effort; however there are a growing percentage of applications and services that would benefit from avoiding the limitations that come with Best Effort. Currently it is estimated that less than 5% of operator traffic fall into this category. However, given the opportunity, this has the potential to grow by resolving the issues that exist today (e.g. lag on networked gaming) and through a new wave of innovation that can begin to occur when a reliable infrastructure across the Internet is possible. Another example could come from the traditional ISPs as they look to move their products from on-net to off-net, and decide to take on the Web Platform companies that have cornered the Internet as a worldwide market. It may also be of interest for the Web Platform companies to be able to interwork with the network operators, as this offers new business model opportunities for them to grow their profit margins.

As the new protocols and functional systems described in the SDN Strategic Architecture become available in the market, flows can be identified and influenced by SDN network and policy controllers for the delivery of enhanced service control.

For these multi terabit per second (and growing) networks, only a small percentage of traffic requires special treatment. This traffic crosses the Web Platform and ISPs backbone/access networks.

SDNs, along with new protocols such as the draft NSH and IP Meta Data, give the ability to extract the relevant flows from the rat nests of active flows, therefore ensuring that the SLAs are met for these flows.

Key to any interworking on the Internet is the concept of peering.

Peering: Many people have many definitions of what this peering actually is. For this reason, the following definition of peering is shared:
Peering is when two (or more) organisations interconnect and share their IP routes across a physical link between both their infrastructures. Using this interconnection, they permit the traffic they generate to traverse their shared interconnection.

Peering ensures that the user (or customer) has connectivity to access destinations on the Internet. To achieve access to all routes on the Internet in a scalable manner, this is implemented using two approaches: Public Peering and Private Peering.

Public Peering happens at an Internet exchange, where a third party provides the ability to interconnect to any other organisation via BGP. Payment methods differ, but the usual method is based on the number and speed of ports used. This is very convenient for many operators who are attempting to connect too many other parties. Public Internet exchanges tend to be commercial companies, and they usually charge each party who is located at their site for each port at a monthly reoccurring fee.

Private Peering is usually carried out between two network operators when the traffic load is significant and in balance (when both parties exchange similar levels of traffic or traffic is exchanged according to an agreed ratio). Using private peering is more cost effective to implement at larger traffic volumes.

Transit is not peering. Transit achieves access to all other routes on the Internet through agreement of a commercial contract. This is used to avoid the purchasing organisation from having to build a worldwide network to directly connect to every other party or ISP on the planet. Companies who offer these commercial transit services usually invest heavily and build worldwide or continent-wide networks to enable themselves to sell such a service. These organisations are commercial organisations; the business model they operate is based upon receiving revenue for the supply of the connectivity to routes that the ISP cannot get direct access to.

Transit traffic is passed (transited) via the supplier network to the receiver network. The receiver pays according to the business model the parties have agreed to, thus avoiding having to build a worldwide network, which is not commercially viable for almost all ISP network operators.

Net neutrality

For most governments, Net Neutrality is about enabling and protecting innovation. The regulation has been put in place to ensure that future business, industries and jobs can be created and sustained in their country through access to an open Internet. These regulations are in place to ensure that companies and individuals will be able to connect to and utilise the Internet. Net Neutrality regulations ensure that the ISP cannot constrain this growth by defining the rules for connectivity on their infrastructure.

In most legislation today, Net Neutrality is focused on not allowing the ISP to set bulk preferences for Quality of Experience or lack of Quality of Experience between itself and another party. In most countries, Net Neutrality legislation permits the consumer to request a service to be delivered with quality. Current Quality of Service technology capabilities limit the ability of operations to simply manage or operate such solutions on an Internet level. In addition, as most people in the industry are aware, the IP protocol is by its architecture an unreliable protocol. Therefore the IP Protocol and Best Effort will be limited in their ability to deliver some innovative high-end services that are being created as new products, for possible delivery over the Internet – in particular when there is a requirement to interconnect to another network operator.

Considering a move to beyond Best Effort

Today traffic on the Internet is bulk forwarded, thus causing a lack of granular visibility of the flows used to deliver services. This lack of granularity introduces limitations that constrain many parties' abilities to identify and manage selected individual flows. It is the individual flows that deliver the user experience to the consumer, and that enable them to enjoy the service they are receiving. To move beyond Best Effort, and to support new business models, visibility is required to ensure the service is delivered to the consumer in the appropriate fashion across multiple ISP networks. For this to work, it needs to be cost effective and automatically programmable to be viable.

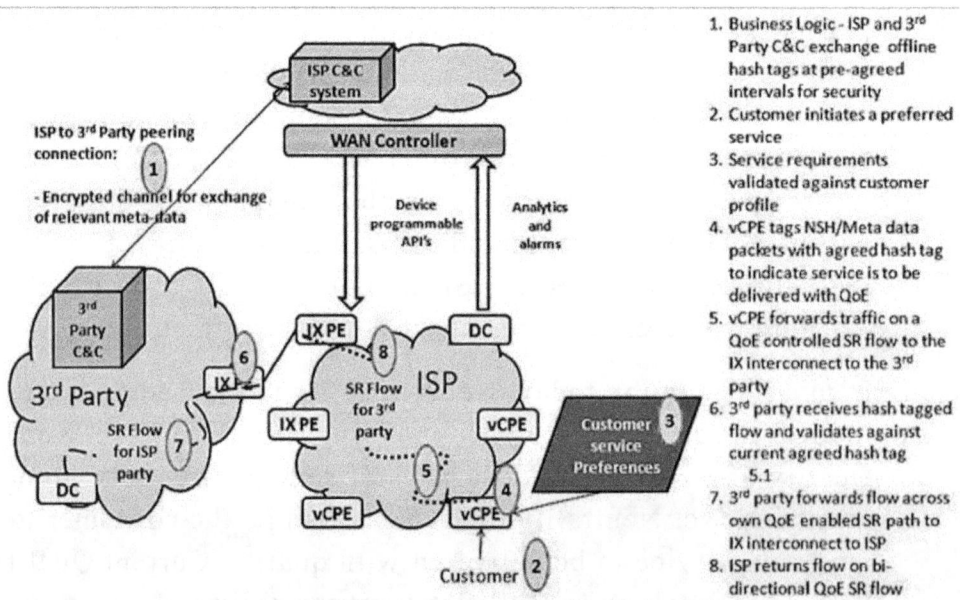

1. Business Logic - ISP and 3rd Party C&C exchange offline hash tags at pre-agreed intervals for security
2. Customer initiates a preferred service
3. Service requirements validated against customer profile
4. vCPE tags NSH/Meta data packets with agreed hash tag to indicate service is to be delivered with QoE
5. vCPE forwards traffic on a QoE controlled SR flow to the IX interconnect to the 3rd party
6. 3rd party receives hash tagged flow and validates against current agreed hash tag 5.1
7. 3rd party forwards flow across own QoE enabled SR path to IX interconnect to ISP
8. ISP returns flow on bi-directional QoE SR flow

Figure 16: Dynamic interworking between ISPs

114

A solution to this situation becomes feasible if on the customer service portal the subscriber could tag certain services (i.e. video or gaming) with their request to have this service delivered with the required Quality of Experience. If this has occurred, the initiation packets would be tagged at the vCPE function when a subscriber triggers one of these services. The vCPE environment then forwards the selected traffic into the appropriate SR flow that was pre-planned and pre-modelled for that service.

This traffic will have been validated against the subscriber's profile to ensure the customer has purchased a product that entitles them to consume this content item. If appropriate, the traffic will then be handed off to the other party and will be marked with a pre-agreed tag that identifies the consumer's expectations to the 3rd party. These tags will be included in the NSH/IP Meta Data context fields and could be pre-arranged and pre-shared on an out-of-band link between both parties. Tags will have different meanings that will be regularly reset, ensuring the level of security that both parties feel is required.

Before crossing the boundary, the NSH/IP Meta Data information will be reviewed to prevent any unnecessary information shared between the interconnecting parties.

The returning service stream will include IP Meta Data information about the content that is required to help guarantee the Quality of Experience for the consumer. This will include information such as the total time of the movie, the size of the stream, the destination for delivery etc. This information can then be used to pre-plan and pre-schedule the capacity on the access network, thus ensuring the customer receives the quality of experience required to enjoy the service. This also gives the network operators the ability to utilise information to proactively manage the service. This permits for individual flows to be extracted from the bulk forwarding traffic and to be granularly addressed.

8.9 Architectural Security Enhancements

Security on a network protects the user and the system by making sure that only those who have the right to access it actually can access it. The implementation of security rules and systems is very complex but this complexity is insignificant when compared to the difficulty that occurs when attempting to validate that the desired results are being maintained and are in place. This complexity is caused by many day-to-day situations: Bugs in systems, scaling problems, lack of visibility, people leaving jobs, processes not being followed, new change requests, the overall intricacy of the solution etc., and in turn these situations create an overall lack of known and provable control.

With this complexity and the cost associated with it, many companies do not have the capability to ensure the correct levels of security, and therefore expose themselves (and Internet users) to intrusion and unwanted monitoring

The SDN Strategic Architecture creates a set of building blocks upon which additional security controls can be delivered. As this is an architecture-based security approach rather than an appliance-based security approach, (appliances are used within the architecture), this permits for a greater number of sources to be used in real-time to validate the access control.

Network operators require dynamic real-time business logic to be fully incorporated into the process of security decision-making. This is required because at the top level it is business policy, business services and legal obligation that define the key business rules for the application of security. The SDN Strategic Architecture supports this subtle shift by putting the focus on network management through the shift from reactive and manual to proactive and automated, by focussing on control of the flow through the network, by delivering the ability to validate the service against integrated real-time OSS/BSS customer data, by policy controlled automated configuration changes to network elements, by using analyse of detailed analytics gathered using APIs, service chaining and by introduction of modelling tools to compute best solutions and practises. Many of these tools have not been available to architects, so this gives the network operators a new business-focused approach on how manage security.

In addition, it is suggested that the service-focused OSS layer integrates a new security function into its structure: For the control of the service security, and for the inclusion and validation of these services against business policy and legal obligation.

The building blocks include the following:

Function Virtualisation: Current appliance-based security systems require that the traffic is brought to them for analysis. With NFV, the security applications can be brought to the traffic. This is standardised and delivered through the ETSI NFV group. NFV brings the capability to spin up virtualised security functions at the relevant point in the infrastructure, to use service chaining to override the destination routing, and to ensure that the traffic passes through the relevant functions for examination.

Policy Control and Network Controllers: Policy controllers provide the ability to deliver a validated configuration automatically onto the network element. This ensures the network operator management process can automatically change the configuration depending on the situation, and allows it to perform checks on network elements to validate that they are correctly configured. If it is found to be non-compliant, the SDN policy and network controllers can be set to automatically re-configure the device or service.

Security function in the OSS: Through enabling an additional security function at the OSS layer, this would offer the opportunity to evaluate the service to ensure that it is in line with the network operator's general business security policy, with the security of the service and the legal obligation the network operator is expected to adhere to. This provides the possibility to ensure policies, applied via the SDN policy controller, are validated in real-time and ensured to be in line with the needs of the business and the consumer. As policies can be structured for reuse, much as is done in an object-oriented approach, this allows for personalised policies to be initiated from within a building blocks structure.

vCPE: With a focus on the consumer, application based products with personal preferences can be enabled within the customer's profile in the Cloud. This ensures that services can be enabled to control and enforce security and to validate that this is honoured. For vCPE solutions, a policy can now be implemented on a virtualised function instance that is relevant to the specific business or residential customer. A policy can be set to be in line with the customer's personal preferences. Compliancy to this policy can then be regularly validated, thus permitting for identification of possible successful hacks and allowing for proactive real-time service security compliance.

Service chaining: The GRT defines the route that traffic will take between the source and destination. This limits the ability of the architect to traverse security functions for flow analysis. Service chaining uses the Network Service Header protocol to override the GRT and permits for a predetermined path to be identified for traffic to traverse between functional systems. When full control checks have been performed, the traffic follows the global routing table to its destination.

Proactive Network Management: Having the opportunity to gather and analyse more analytics provides new opportunities to address security issues. New security analysis applications can be created that can identify issues and can be hooked into the policy controller to affect solutions provisioned into the network. Proactive security management applications will need to be developed to provide the capability to generate a new approach to network management.

8.10 Summary

As can be seen from the approaches described, SDN is not the only technology evolution underway. It is, however, a key evolution that

118

embraces change and enables a new approach to be considered going forward. This change drives simpler development of the solutions, which may be immediately considered by some as a threat to their jobs. This need not be the case, as the shift from infrastructure control to service management provides the opportunity for companies to create new products and in this process creates greater opportunity for those with the skills and knowledge. With those in the lead being able to define and influence the directions projects take and the products that can be created, it is now up to the technologists to advise the product teams of what they should be defining and what is possible.

The solutions that were discussed indicate the level of completeness of the thinking that SDN has generated, and how the solutions that are developed have an end-to-end view on what the network operator requires. This shift from focussing on the assembly of tightly integrated components to focussing on a platform that provides a solution, now starts to fit with the work that an engineer or an architect have been delivering.

9. SDN: Service & Infrastructure Management

As an industry, we are all fully aware of the complexity involved in bringing together the required components and protocols to create an interworking service solution. At times, the result is mind-bogglingly creative and can almost be considered a work of art. However, this creativity is sometimes not fully appreciated when handed over to the operations teams, and the comments that are received put somewhat of a downer on the months or years of work. The operations team's reaction to the handover of a project is naturally depending on how complete the operational handover process is, and when lacking, the operational teams, rightly so, see it as a solution that is impossible to manage and operate for the customer. The difference in perception comes from one group having had to examine a very significant number of options and combinations to create the solution, versus the other team now knowing they will have to operationally manage a very complex solution on a daily basis with limited management insights and tools that are not proactive by design.

This operational complexity occurs, partly because of the limitations caused by component-based architectures and because of the lack of cross-vendor infrastructure and service management technology. For the network operators and other parties involved in the creation of SDN, one of their founding principles has been to address infrastructure and service management. This is because the historical component-based approach did not fully address the management aspects of solutions creations that are a priority for all network operators. Focus on management and control has always been lacking in comparison to the forwarding plane capabilities delivered for networking solutions. This lack of multi-vendor focus on management has significantly driven up the OPEX costs and has severely hindered the deployment of many projects for network operators.

SDN is different to the historical component-based architectural model, in that, SDN incorporates the principles of delivering service and infrastructure management as primary goals of its architectural approach. To drive this technology principle, building blocks have been enabled within SDN and the SDN Strategic Architecture to deliver enhanced management capabilities. These developments extend beyond management of the network infrastructure and into direct service management and control.

This chapter aims to provide an overview on the new management tools that are becoming available within the building block toolbox. As this book focuses on the strategic architecture, this chapter also discusses some additional functional management systems that could be developed to further enhance management within an SDN Business-Focused Strategic Architecture.

Factors that have influenced change

Change is not being driven because of SDN. Change is being driven by many network operators because of the immense complexity of networks, the complexity of managing service delivery, and because they are struggling to control and manage their environments accurately. Therefore, one of the reasons that SDN has been developed and sponsored by network operators is as a reaction to the lack of management capabilities within the current technology model required to be used by network operators. This is because the lack of end-to-end control of the network infrastructure and service, limits the ability of the network operator to drive new revenue models, and because it reduces the completeness of the services that a customer can receive.

Cloud and SDN have a focus on the management and control of the service and aim to utilise the work done through NFV to achieve this. This new spotlight on management as a primary focal area of the technology development permits a major shift: From managing services through focussing on controlling the network infrastructure to managing services through managing the flows generated to deliver the service.

This strategic technology change is made possible through SDN with its introduction of APIs into the network elements, the network and policy controllers, and because of new developments in the IP protocol stack with the creation of new protocols such as NSH, IP Meta Data and SR.

These changes identify the shortcomings of the current OSS and NMS systems and identify an architectural approach to be considered. SDN, NFV and Cloud define a new approach to how services can be created and how networking infrastructures can be utilised and controlled, and with this fixes many of the limitations caused by the historic networking approach. SDN permits for the exposure of data about the service and infrastructure. With a new level of analytics data now capable of being exposed from the network, the OSS and NMS systems can now be redefined to utilise this information and to therefore better fulfil their role and function within the network operator organisation.

Historically due to the lack of openness in networking equipment, the OSS and NMS systems were only able to attempt to find out what might be the state of the network, rather than to be informed of what the state of the network or service actually is. Access to new levels of management data through the SDN architectural approach now allows for the creation of new network management applications that can utilise the API communicated data. This change in approach now permits the shift from reactive to proactive network and service management. To fully realise this, a review of the existing network management architectural approach is required, and a new strategy needs to be defined. This chapter considers some systems that could be utilised within such a strategic architecture.

Historic network management problems

Networks have traditionally been managed by using alarm and event notification NMS solutions. These tend to use a traffic light system, highlighting problems for the NOC staff to investigate. Thus the goal of many NMS systems was to notify a human in the NOC that something had gone wrong, and to alert them that they need to identify the reason for the problem.

To investigate further after receiving an alarm or events, the NOC staff - applying personal knowledge and manual processes - are required to access the numerous proprietary element management systems, scripts, databases, open source service or infrastructure tools, proprietary service or inventory tools – anything that can help derive the necessary information to resolve the problem.

The initial investigation path followed by a NOC engineer is usually dependent on which alarms are received first, which team in the NOC received it, and whether or not the alarm was appropriately configured.

Depending on these and many other factors, the NOC engineers can be sent off on a wild goose chase because of the lack of accurate multi-layer correlation of faults. This all leads to service and infrastructure downtime, customer dis-satisfaction, loss of sales, loss of contracts, greater workload because of new tickets being raised due to related but uncorrelated events, etc. and can trigger payments to be made to customers because of SLA failure.

NMS, like OSS, has relied upon open standards to evaluate what is happening on the network. Protocols such as SNMP, the proprietary predecessors to IPFIX, syslog etc. are extremely basic tools for a large and complex industry. Some vendor proprietary element management solutions do give a good insight into what was happening on a closed section of the network infrastructure, but none have been successful in delivering an end-to-end multi-vendor and multi-layer view.

This complexity has been further enhanced by the many variants of vendors' products, generations of products, variations in command line, the command line itself, etc. This multitude of vendors and their products requires a large number of screens to display the events and alarms, and for most network operators this requires a Manager-Of-Manager (MOM) architectural solution.

Due to the large number of non-integrated management systems and the lack of multi-layer correlation, these systems can create workflow process flows of varying quality and tend to generate a significant amount of work for the teams involved. In addition, staff turnover tends to be higher in the NOC teams as this is considered by some as a lower skilled job.

As with all organisations where staff turnover is high, historic knowledge and understanding of the solutions in place are soon lost. This is solved in some organisations as they rethink the role and function of the operations teams. The organisations are moving away from the old NOC model and are changing the functions and expectations of the staff by running within a DevOps model. This model empowers the staff and helps attract skilled engineers who previously would have expected to take on engineering roles.

SDN and the SDN Strategic architecture fits with the thinking of the DevOps model as it enables faster implementations, greater control, automates fault resolution and gives the staff the tools they require to deliver according to the needs of the business.

Industry Change

Historically, many network operators focused on a single access technology, and common terms were used around the industry that reflected the business model these companies used. With recent acquisitions and new network builds, network operators are now moving from their historic singular focus to having to support multiple access technologies.

As these business models change, the skills required by the NOC teams change with the increasing number of technologies they will have to comprehend and to resolve issues for. When complexity occurs, the opportunity increases for customer impact.

This re-focussing of the industry drives the need for a new architectural approach as the increasing complexity requires a management solution that is multi-access technology and automated. Network management solutions cannot provide this by themselves, as this requires an end-to-end approach, from development through to delivery and management. The SDN strategic architecture sets out an end-to-end approach across these separate domains.

124

9.1 SDN Strategic Mgmt. Systems Structure

SDN brings a clear focus on service management. With the separation of the network control and forwarding plane functionalities and the introduction of the central controller, it is now feasible to interface service and infrastructure management applications to create a new focus automation of end-to-end management. The old adage that there is no point in designing something that cannot be managed and billed for is now given a full focus. The goal of the SDN Strategic Architecture is that this old adage also gets changed. This new adage aims to be, there is no point in designing something that cannot be "automatically" managed and billed for.

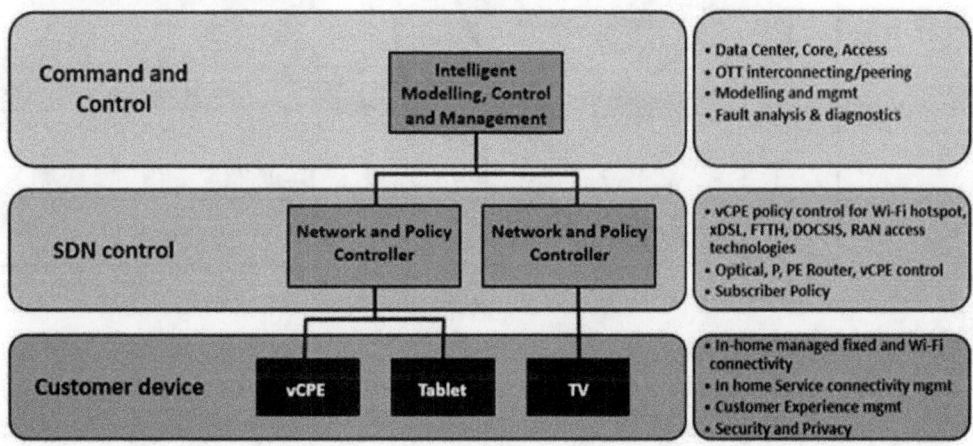

- **SDN delivers the integration of network infrastructure control and infrastructure mgmt.**

- **SDN permits end-to-end control of the service across multiple historical silo's**

Figure 17: SDN Strategic Architecture - Service focussed management

In an SDN topology, the network and policy controllers provide a central control point from where policy-based actions can be triggered to control and optimise the network and services. Through the controllers the network administrators, management platforms or portals have the ability to create and automatically push control changes which enable a change in the forwarding plane.

The SDN Strategic Architecture introduces the concept of a Command and Control (C&C) management and governance system for the end-to-end environment. The C&C system is a term that has been conceived to describe all the systems involved in management, and it aims to deliver a management architectural concept which enables end-to-end service management and control.

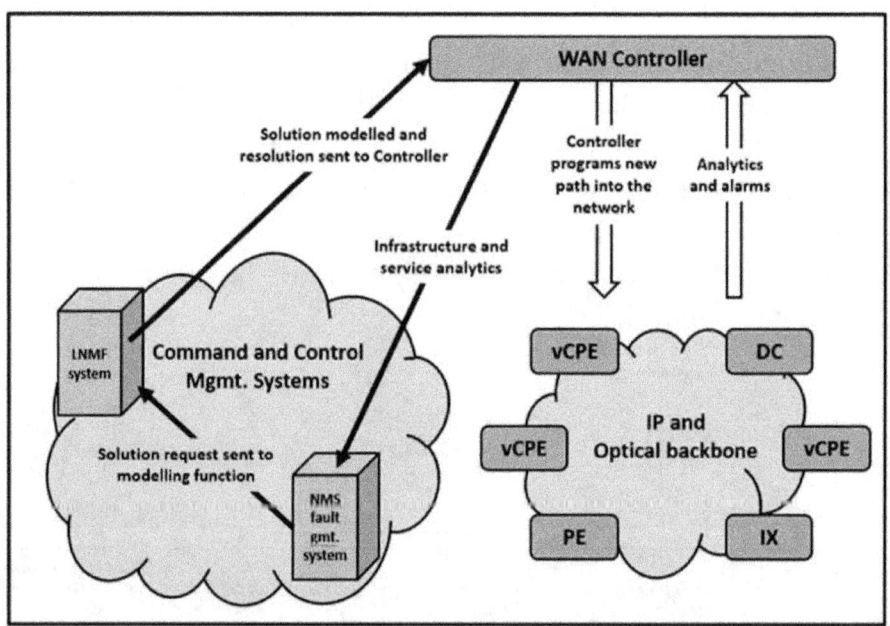

Figure 18: Automating Network Management

This will be delivered across the multiple layers of a network, connecting the OSS and BSS systems. This permits for the inclusion of business logic and business policy into the decision-making for the control of services. This is provided through applications that reside as part of the centralised and distributed management and control functions within the architecture. These capabilities provide control and management of the individual and aggregated flows across the operator's network and the Internet. Some of the suggested architectural systems that are described within the C&C are based upon functional systems that are under development; some still need to be developed. When these systems are interfaced together through APIs, they offer the opportunity to enhance the experience of the user through a focus on service delivery and with the aim of delivering automation within the network.

126

Without near real-time or real-time data a service can't be controlled. APIs introduced into networking by SDN give the management systems the visibility of key analytics that influence the service delivery and identify data relevant to infrastructure control.

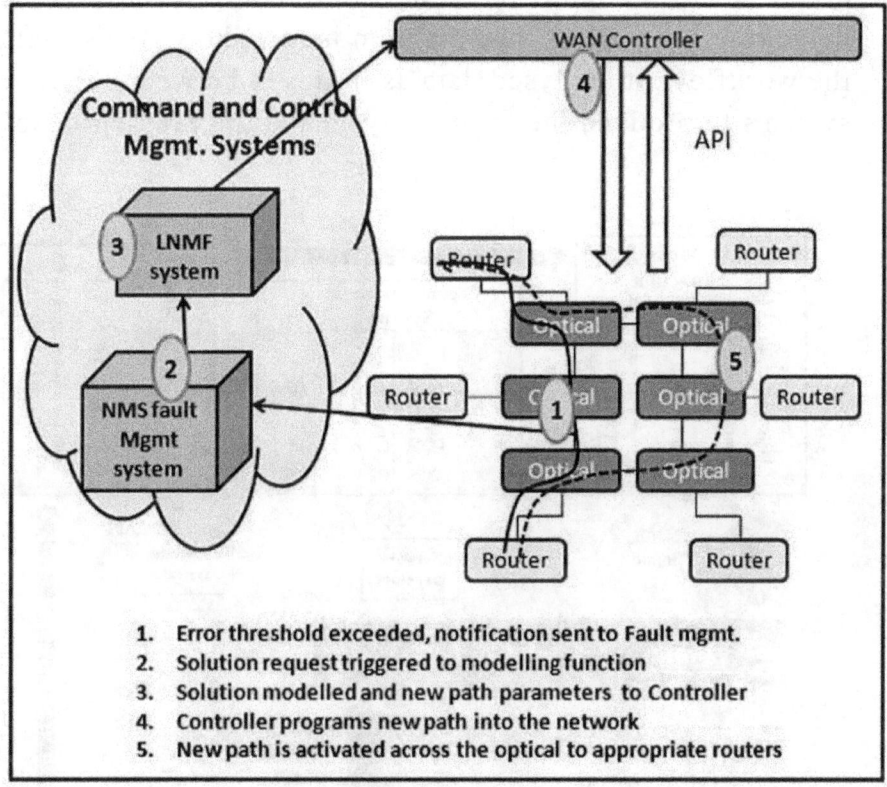

1. Error threshold exceeded, notification sent to Fault mgmt.
2. Solution request triggered to modelling function
3. Solution modelled and new path parameters to Controller
4. Controller programs new path into the network
5. New path is activated across the optical to appropriate routers

Figure 19: Automated Optical path fault resolution

Addressing the end-to-end architecture and enabling new functions brings the influencers of service failure into visibility and brings to bear a controllable end-to-end solutions architecture.

In terms of how the business operates, SDN probably has almost the same impact on the NMS and the OSS stack as it has on control of the networking infrastructure. SDN changes networking from being a fully distributed to a centralised solution with a distributed influencing layer. In achieving the centralised view, it attains a reference point from which it can view and influence the network in its entirety.

9.2   SDN Functional Management Systems

The toolbox identified in the SDN Strategic Architecture provides the ability to move to an intelligent management system that can drive the move from reactive network management to proactive network management. The diagram below shows the structure and the workflow of analysed data as it moves between the functional systems to produce the fault analysis and the problem resolution.

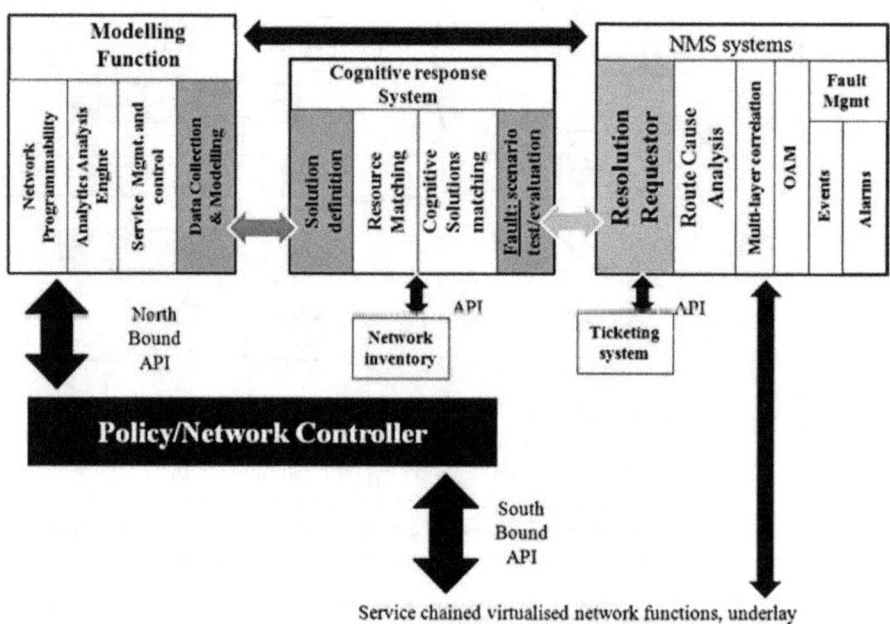

Figure 20: SDN Command and Control Solution

This functional structure is identified to provide a connectivity matrix for the production of automated fault resolution. It should be understood that the following list of functions is not definitive. As always, individual business models used by specific network operators will demand niche solutions. These niche solutions will require that the engineers and architects identify and define additional applications to address those requirements. Such niche solutions can be created through internal development, shared open source, or through vendor partnerships. The development model selected to create these applications could be based upon the financial return they could bring the network operator with customer support and OPEX control.

The purpose of this structure is to set the strategic goal of reaching network self-management, and to ensure a path that caters for controlling OPEX and achieving real-time service management. To attain this, some new functional systems are required within the network management architecture to enable intelligent decision-making. This sets in motion a strategic path towards accurate fault resolution.

SDN Fault Management system

Fault management systems collect and correlate notifications from the network elements.

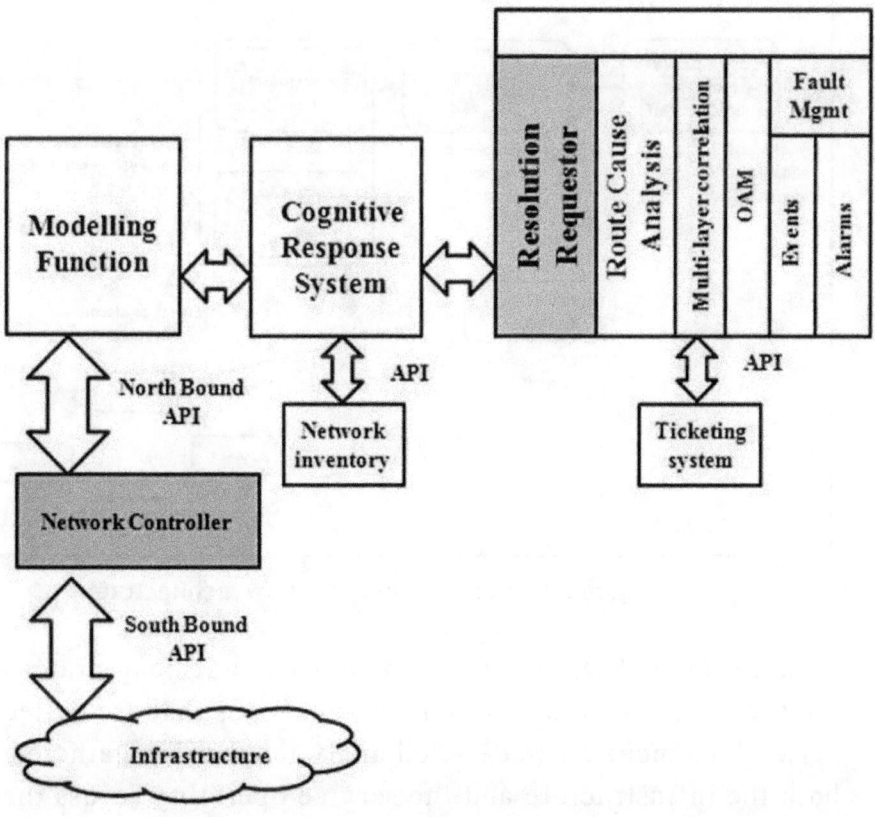

Figure 21: Fault Management functional components

Depending on the category of the alarm or event, the fault management system highlights the appropriate error message to the NOC team.

Within an SDN networking infrastructure, today's usual fault management systems continue to be retained and used. However, a strategic architecture now requires that additional capabilities in the SDN controller application layer are provided for fault analysis and correction. This ensures that faults for which use cases are understood and defined can be addressed in real-time. The number of faults able to be addressed will grow over time as the use cases are evaluated and shared.

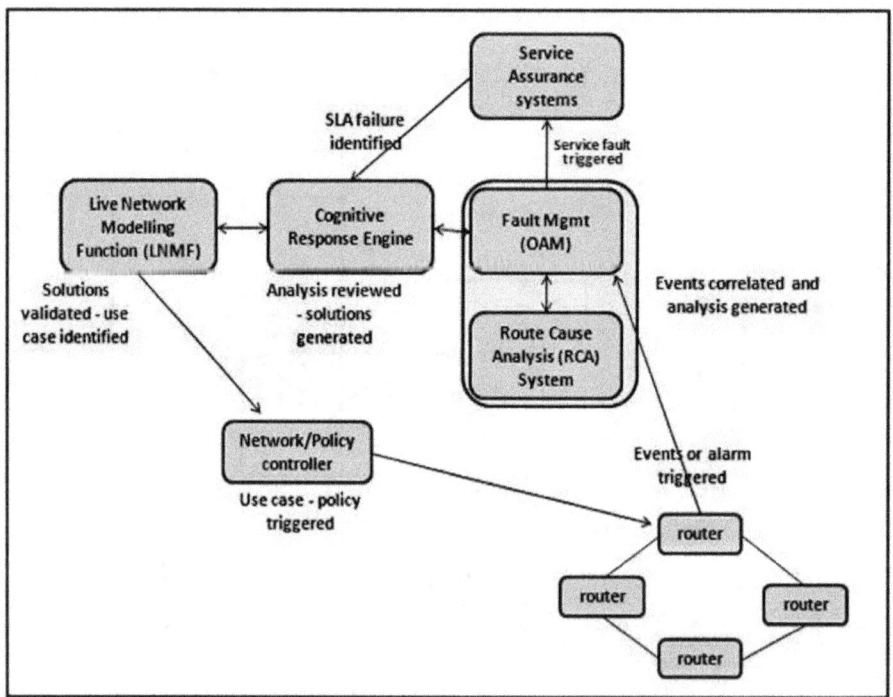

Figure 22: C&C: Strategic NMS flow architecture

In addition, an SDN-focused network management strategy requires the processing of the detailed data analytics that are collected from network elements. The detailed analytics that are gathered concern both the infrastructure and the service operating across the network.

This is collected using the APIs by the controllers provided within the SDN architecture from the network elements for both overlay and underlay related data. These detailed analytics can then be processed using the cognitive response system against known use cases to best identify the source of the problem.

Cognitive Response Function

This description is not exhaustive, and serves as a proposal to permit the reader to consider different scenarios and use cases. This solution would be expected to reside between the NMS fault function and the LNMF. Via an API it would hook to the Network inventory solution to permit validation of resources.

As the SDN Strategic Architecture incorporates the subscriber profile into the networking environment, business logic is introduced into network management logic. This enables a fault resolution approach that supports the needs of the business. With the SDN controller and the SDN Controller applications incorporating real-time information from various sources, the opportunity for an RCA (Root Cause Analysis) application is created. In a next step, this could be used as part of the operational process to support the engineers and as a trigger for automation of ticket creation. The ticket would be populated with the synopsis of the fault and could identify the likely network areas that had caused the issue. As the SDN Strategic Architecture integrates the subscriber profiles, this would allow for an automatic correlation to be run against the likely affected subscribers when a fault occurs – which would allow for a service priority instead of only a technology priority to be set for the ticket. The NOC staff could focus on the business-affecting tickets, as the true priority of the ticket for the customers would have been identified.

These capabilities become possible because of the centralised view the controllers and their associated applications provide due to an SDN controlled network. These capabilities can be further amplified through the introduction of the LNMF modelling capabilities, creating the opportunity to enhance the multi-layer correlation capabilities beyond just the automatic definition of a RCA (Root Cause Analysis) report.

A cognitive response system could be created to identify the correct solution to a diagnosed issue. This could initiate a request to the LMNF for it to model and validate a selected resolution that the cognitive response system has identified against the possible impact to a live network. When the network operator processes have been completed, the SDN controller could be used to program the solution into the network.

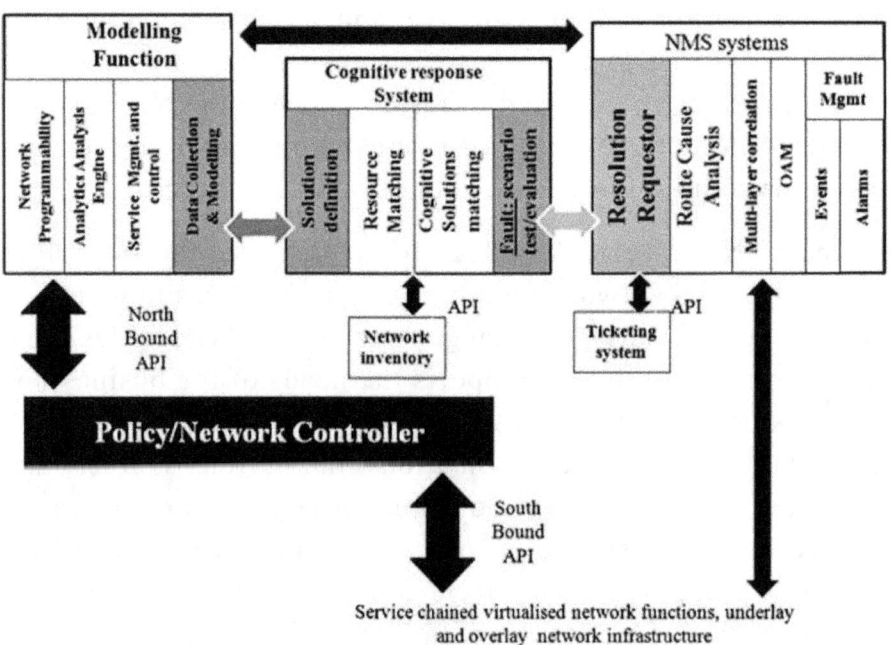

Figure 23: C&C Overview: Strategic NMS Architecture

The scenario use cases utilised by these solutions would first have to be identified, templates would have to be created, tested and pre-modelled by the engineers. This approach fits well with a DevOps environment. This would allow for the automatic enforcement of corrective network management capabilities to be implemented, thus permitting for solutions to live problems to be invoked at a granular level. Solutions will require interfacing to the ticketing system and the resource management system.

Additionally, this proposed architectural strategy requires the development of a resolution requestor application based upon the correlated outputs gathered from the network from the NMS system. This will be transmitted to the Cognitive Response Function via an API, where this information would be analysed, evaluated and

132

tested. It may at any point trigger an analysis on the LNMF. Possible solutions will be matched in the cognitive solutions matching function; in a next step resources will be queried against the resource management system.

The selected solution will then be defined, and a full analysis of all impacted endpoints will be correlated. The general programming requirements will be identified, and an appropriate policy will be defined to affect the fault resolution. This will be communicated to the LNMF that will run its final validation of the solution. This may then activate the change on the network automatically, or may first be validated by an engineer. The solution would then be programmed to the network to resolve and/or to address issues on the environment.

The modelling function (LNMF) is now under development by most vendors, but the cognitive response system is not getting the same functional level of development success. Such a functional component could be a key function for enabling automated management of networks. This function is a key one for architects and designers to drive development of against their fault and problem service and infrastructure use cases. Development should be delivered within an Object Orientated architectural approach, as this will ensure the reuse of functionality developed for existing use cases. This solution can also be expanded beyond fault management and considered for use for Peering analysis, Off-Net service capability analysis, pre-positioning of service configuration, resource confirmation, and others.

Architecturally, these new functions are best introduced as new functional blocks into the NMS architecture. Introducing a new specialised function system ensures that these capabilities are not introduced in existing systems, as vendors try to enhance their solutions to differentiate themselves. A functional block solution is required in the NMS architecture to ensure future openness of the NMS architecture. If this capability is simply built into an existing suite of products, operators will find themselves presented with a multitude of systems with overlapping and comparative functionality, which brings about a highly confused architectural model.

Resource and analytics management

Key to controlling a network is the ability to manage the resources that are required to identify the services utilised in the network. For this purpose the management of resources is required. Resource management is a generalised term that covers the various systems listed below. This centralised resource management capability is connected to the policy controller, and template-based configurations are automatically allocated to the next available resources for locally and globally significant resources.

The **topological database** maintains a record of the active state of all the network resources. This database records the multi-layer network capabilities; this includes visibility of the underlay (physical optical, IP, Ethernet and other infrastructures) and the overlay (logical assigned capabilities), topologies and the resource available or allocated.

Real-time accounting records statistics covering all layers of the topology which are relevant for day-to-day management, such as throughput, OAM, and others.

Resource management records and tracks infrastructure and service layer resource utilisation across all layers of the network. These records are used to support other Command and Control queries for the positioning of new services and to support the generation of LNMF function queries.

Tenant Management controls the multi-tenant authentication and access rights management to network resources. This function is required in the network management system to ensure that resources are identified and can be delivered when a service requires them.

Network Services Inventory maintains the real-time state of the networking infrastructure and services to ensure that an accurate reference system exists. This is critical to ensuring services are delivered at a granular level, and that any constraints on resources are known should a new service be requested. This data permits for the introduction of proactive services and infrastructure management and Root Cause Analysis generation. These are all building blocks that enable and ensure that the customer receives the service in line with their expectations. These solutions with their enhanced monitoring capabilities enable this change in the approach to network monitoring.

Live Network Modelling Function (LNMF)

The network modelling function (LNMF) is not new in IP Networking, as these solutions have existed and have been used in operator networks for more than a decade. These are sometimes referred to as modelling systems. Historically, they have been offline systems that have been used to support the design work, capacity planning, costing and budget support, and to enable operational stability on the networks by advising on better parameters to use to program the network.

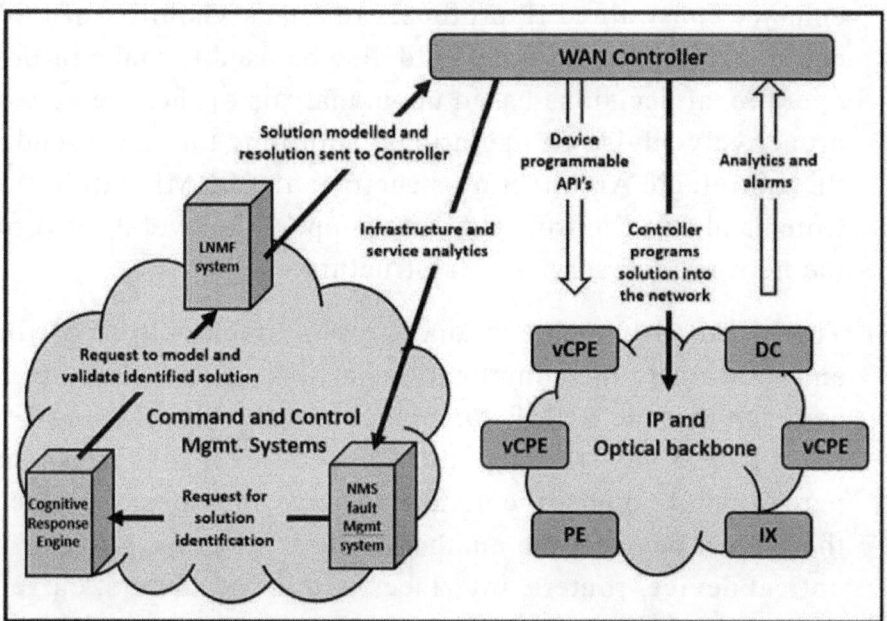

Figure 24: Management through Analysis and Modelling intelligence

The extension that the SDN Strategic Architecture brings to these systems is that this function now operates in real-time - and therefore allows for the LNMF to add multi-layer intelligence to management systems, as it considers the multilayer end-to-end solution for operational and design support.

The term "Live Network Modelling Function (LNMF)" has been created by the author to uniquely define a term for the network modelling function. This is to ensure that the networking teams do not use 'modelling' when referring to networking modelling because the CIO teams use the term modelling when referring to data modelling. The term LNMF has been created only to differentiate it from the 'modelling' term that is commonly used by CIO staff. Such use of the same word for very different functions tends to cause inter-departmental confusion.

With the LNMF bringing a multi-layer (layer 0 to layer 4) end-to-end view, this allows it to be used for predictive design analysis, e.g. to understand the impact of new service on the infrastructure, thus enabling a strategic path which gives operational and design value. This approach, when further developed, can drive a reduction in the overall architectural costs of the network and can be used to enhance constrained IP protocols that lack visibility of the end-to-end network. The LNMF would also be used to make informed operational decisions based upon analysis of the live network and to proactively advise on operational solutions for end-to-end. The SDN Strategic Architecture structures the LNMF within the Command and Control (C&C) to support the overall governance of the network and service infrastructure.

The LNMF also is able to analyse many factors in an offline mode and to evaluate their implications against the architecture. It is common that the LNMF can model at a flow level using IPFIX (Netflow, SFlow, JFlow) data, at a duct level for fibres, shared fibre paths in those ducts, the distances between nodes, the placement of the optical devices, the number of cards, the capability of the optical device, routers, interfaces within the routers, current load, and can trend historical load.

In addition with further development content placement could be modelled to limit the impact of network load caused by certain content sources. If a piece of content, distributed using a CDN, was causing load on a constrained section of the network, the LNMF could analyse this and trigger the CDN to distribute the content lower in the network through the spinning up of new vCDN instances.

The LNMF can then factor in the best method to optimise the end-to-end service across the network while, simultaneously pushing new metrics and/or configuration through the policy/network controllers to optimise the traffic on the network, according to the needs of the service or the value of the service to the operator.

The following table describes steps that can be used within a C&C system to utilise the LNMF system to move to proactive network and service management.

Managing the Core		
	Action	Application Function	
Step 1.	Receives reporting and alarming from all devices	Fault Management function	Receives traps and alarms
Step 2.	Request solution	Cognitive Response system	Review symptoms against use cases
Step 3.	Model options and identify solution option	Live Network Modelling Function	Model multi-layer
Step 4.	Solution identification	LNMF/Cognitive Response Systems	Report to Engineering staff solutions
Step 5.	Solution activation	Policy Controller	Implement solution based on automation or Engineer approval

The Live Network Modelling (LNMF) function, with further development would be the first network management system that provides a end-to-end multi-layer capability to intelligently analyse and advise on situations within the network. With this system's ability to compute a vast array of variables, this drives a strategic architecture that can identify incidents and faults, and push solutions into the network using the policy/network controller function. With the Internet becoming more crucial for business, this form of creative intelligence is needed to support the network and the staff running the organisation.

The LNMF system can be used for analysis and solution identification of multi-layer capabilities such as optical, and with the advent of segment routing, IP path programmability as well. Segment routing is used to aggregate flows that have the same sensitivity requirements on the network.

When it is identified that trends are changing on the network and that SLAs may be impacted due to an outage, a node failure, growing load on part of the path, etc., the LNMF – upon receiving analytics from IPFIX and others - can evaluate the changes on the network and push a solution, therefore enforcing proactive corrective action. Through addressing problems proactively through analytics, faults are avoided rather than having to be addressed when the problem has progressed to impacting services.

The analytics for this capability will be further enhanced when Network Service Headers (NSH) comes into common use. NSH supports per hop reporting using the OAM capability in its structure to communicate the data indicating any changing patterns on the network to the LNMF. With this level of visibility, the LNMF can take on the role of proactive network management.

Policy Controlled Customer experience

Policy-based customer experience is not the same as the policy control of the network infrastructure. These are two separate and very different functions. Policy-based customer configuration refers only to the configuring of the product that the customer has purchased, and the possibility for the customer to have features enabled specific to their customised personal preferences.

This capability is developed to be included in the vCPE solutions with the purpose of ensuring that the customer's needs are addressed in real-time and in line with the product they have purchased from the ISP.

The needs of the customer are changing and there is a wish for them to have personalisation within the products they use. This is a significant move away from the product boot file approach of today - a few products for millions of customer - to attaining a policy-based customer product that allows for millions of products for a few customers.

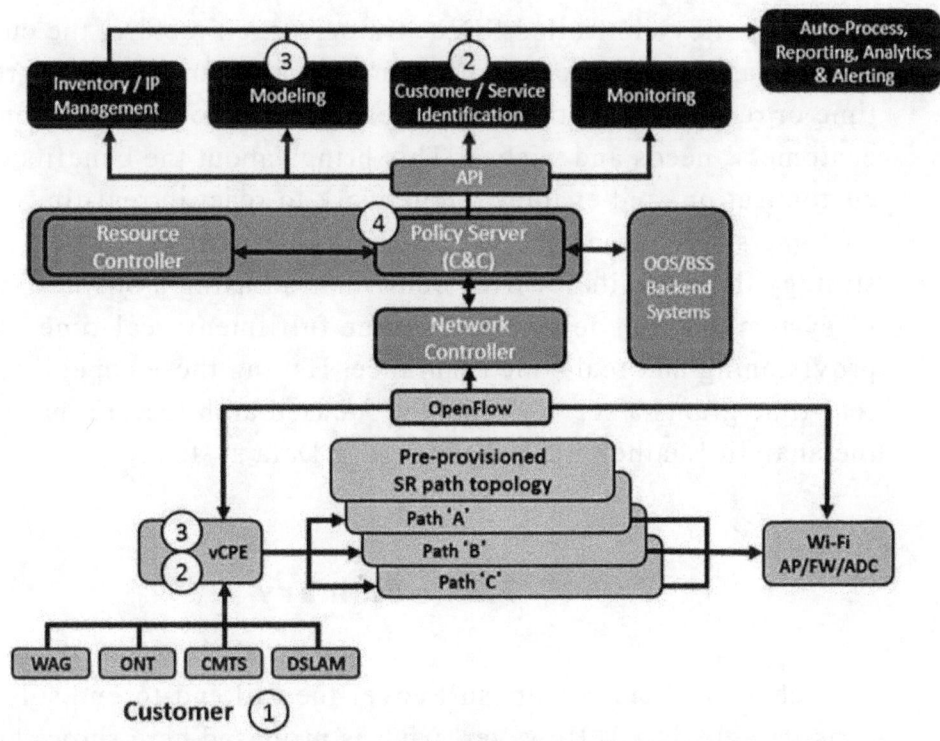

Figure 25: Customer SDN policy controlled services

With policy-based customer products there will be greater flexibility in creating products that support the needs of the end

139

user. This can and will include better control of security and privacy while on-net, simplified configuration, a greater number of applications and application checkers, and other benefits.

This way, customer preferences can be indicated, which permits different responses depending on the service in use, the user, the time of day, the quality required, and other factors.

Integration of the local vCPE policy-based controller with the SDN network controller will permit broadband service management for fine-grained control of subscriber traffic flows. This will also allow for the integration of SON type access technology management to ensure real-time reactions to the network state and the proper positioning of the traffic into the correct Segment Routed flows to ensure automated control and management of the service the customer instantiates. Essentially, this permits for on-the-fly adjustments depending on the controls required for the service. Using a policy-controlled SDN Strategic Architecture, the customer profile can be referenced, which enables an immediate near real-time or real-time reaction to be created for a flow based upon the customers' needs and wishes. This brings about the benefits of customisation, and enables the network to react in real-time to the customers' selections and choices. These capabilities foster a strategy that permits for SLA confirmation using a business-focused IT system that can deliver a real-time fulfilment, real-time provisioning and real-time assurance. Having these capabilities in real-time grants a service delivery focused architecture that utilises the analytics gathered through the Big Data system.

9.3 Summary

This chapter obviously doesn't cover the full end-to-end solution in considerable detail. However, what is proposed here shows how SDN, through its separation of control and forwarding, has created the opportunity for a new approach to advanced service and infrastructure management.

Even at this early stage of development, SDN solutions already available on the market allow for the solving of some of the repetitive problems that consume the time of the operations and engineering teams.

It is accepted that the complete automation of management is probably not feasible, however when the author considers the potential of some of the solutions that the author has already personally witnessed in action during evaluation of some SDN solutions, it makes him wonder as to what two more years of development might bring.

Through this described approach, this proposed strategic architectural approach, with its goals of proactive network and service management and control, sets out a path for new more open solutions to be created. This development can be achieved through vendors, internal teams, Open Source groups or by start-ups.

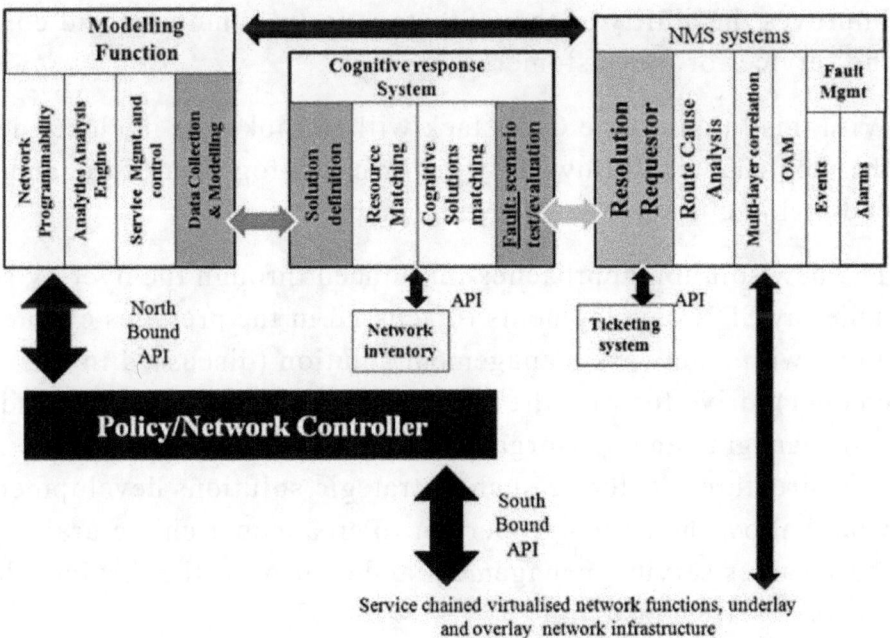

Figure 26: Strategic NMS functional building blocks

The model chosen for product creation will be a business decision that will have to be made by the network operator. Development of these solutions can create the ability to differentiate one network operator from another, and internal development, although not normal for network operators, should not be discounted. It is also hoped that some of ideas will be taken up by readers to create a project or start-up to build such functional components.

These problems need to be attended to and addressed because they are constraining the business model of the operators, and they are also causing unnecessary stress and inconvenience for those working within our industry.

By discussing existing systems, new developments that are coming to the market, and by identifying some new systems, this chapter describes the building blocks sitting between the SDN controlled overlay and underlay networks, and the connection to the OSS stack.

As the OSS stack has the key responsibility of managing and controlling the service for the customer, a proactive and OSS-interconnected network and service management function only improves the ability of the OSS stack to fully monitor and control the service for the customer.

With this in mind, the OSS stack will be looked at in closer detail in the next chapter, followed by the bringing together of the end-to-end architecture in the following chapter.

The new solution approaches introduced through the overlay and underlay SDN developments (discussed in the previous chapter), along with a network management solution (discussed in this chapter), drive forward the concept of achieving a self-expanding, self-managing and self-organising network and service infrastructure. Fostering such a strategic solutions development would allow the network operator to create an architectural roadmap that enables service management and control of the services the customers are consuming.

10. SDN: Network-integrated OSS Architecture

Like many network engineers, I too have for many years considered the OSS stack to be a roadblock to delivery. However, while developing the SDN Strategic Architectural approach, I was faced with the realisation that it is, the networking technologies that constrain the OSS and NMS stacks and caused a significant part of their complexity. The lack of exposure of relevant data from proprietary network solutions, and the proprietary complexity required to trigger a resolution from the OSS stack into the network, has caused both the NMS systems and the OSS systems to grow in complexity and to limit their capabilities. However, this is not to say that all problems in the OSS stack are caused by the network technology stack.

Now that SDN has corrected many of the networks limitations that have constrained the OSS and NMS developments, the OSS can be structured to singularly focus on the business needs (the commercial SLA) of service management and its control. By using APIs within the SDN architecture, enhanced analytics can be made available from the network, therefore permitting Big Data to be fully incorporated into the OSS/NMS management structure.

These combined advancements drive a strategic architectural development that can produce the integrated business-focused architecture that the network operators require, for the delivery of an end-to-end customer-focused service solution.

Competition has grown with new competitors entering the market. These companies offer traditional customers a new approach to gain access to services that once could exclusively be supplied by the PTTs, cable operators or mobile operators. With this, the expectations of the customers have changed, they now expect more from their suppliers: To have products that are in tune with their needs (customisation and service flow management). In addition, the changing business models now raise expectations from senior management for faster turnaround of service development and service delivery.

SDN overcomes many of the limitations of networking, and it offers the OSS stack the opportunity to access the service and infrastructure management data that previously wasn't accessible. This allows for a change of focus for the OSS stack as it is no longer necessary for the OSS system to be limited to trying to manage the infrastructure as a means, of emulating the management of the service. There's space for a shift of focus.

The impact of an SDN/OSS integrated solution

Historically, the service has been managed by trying to manage the infrastructure devices. This has been partially successful but gives limited ability to fully manage and influence converged services delivery in real-time. With network operators converging services onto unified backbones to achieve cost reduction and to permit cohesive communications, the aggregation of IP flows has caused a loss of service control and visibility due to the current limitations in managing and controlling service related IP flows.

During the evolution of OSS design, it has been constrained by the inability of the networking equipment to expose the required detailed service and infrastructure data from the live networking equipment in real-time. SDN networking solutions by exposing detailed service and infrastructure related data, offer a new capability in service management.

A key point to review at this time is that the OSS systems have not historically been able to access the detailed service and infrastructure related data from the networking equipment. Therefore, the current OSS systems architecture and solutions in use today have not been designed to utilise or to consider this data. No matter what architectural approach is considered or used going forward, the OSS systems now need to be strategically changed to utilise this important business related data and to ensure end-to-end service delivery is automated and fully managed.

SDN solutions available in the market not only overcome some of the historic constraints of networking equipment, but now also take on the real-time policy-driven responsibility of management,

144

control and provisioning of these capabilities into the infrastructure. To do this, SDN Controllers solutions integrate some applications that would have historically resided in the OSS Assurance, Provisioning, Orchestration and Fulfilment stack – such as resource management databases, topology maps, and others. The controller uses these applications to structure template-based configurations. These are then triggered through the real-time controller to take corrective action against faults, or to provision new features into the network infrastructure. This produces automation of configuration placement that can advance services and infrastructure management and control.

With Cloud, Big Data and NFV providing integrated functional components and SDN introducing enhancements to the networking environment, e.g. new IP layer protocols, policy controllers, configuration abstractions, multi-vendor support, integrated management systems, modelling functions , etc., the question must be raised: "What should now be the focus of the OSS stack?"

Historically, the OSS stack has had two functions to perform:

This first function has been the structuring, creation and pushing of the technology configurations into the network and servers (including managing resource management etc.).

The second function has been the analysis and management of the service that the customer has purchased. This business delivery analysis has been used to trigger changes to the service and to report to the BSS stack whether the service is fully functioning and compliant to the SLA.

The first function is addressed by SDN and the other technologies, but the second area of focus isn't addressed. This function is crucial to the business, and therefore the OSS stack has an extremely critical function in the network operators' infrastructure. This provides the ability to restructure the historic approach to service management and to concentrate on the customer experience, thus creating an opportunity for the network operator to differentiate themselves within the market.

As many operators look with increasing interest at customisation of customer services, there's a growing need for a more flexible real-time OSS stack. To achieve this capability, some OSS stacks will require restructuring to support the complexities produced when dealing with a customisable product approach. These new requirements now need the CIO and CTO functions to work together and to define how they will deliver this dynamic stack.

Standardisation of a new OSS Architecture

SDN, NFV, and Cloud all drive change into the OSS stack. These changes are recognised across the industry and are highlighted by the level of action that has been initiated. Standards bodies have recognised the impact of SDN on OSS, and work is now progressed within ETSI under the mantel of NFV MANO. This OSS body is mainly driven by operators who recognise the need for a new OSS architectural approach and who take advantage of the capabilities delivered through Cloud, SDN and NFV.

The Next Generation Mobile Networks (NGMN) Alliance, through the work being done by the Next Generation Converged Operating Requirements (NGCOR) group from the ETSI NFV MANO are evaluating a new approach which incorporates the implications of SDN, Cloud and NFV architectures.

SDN/OSS industry cross-over

Through the early development of SDN, Cloud and NFV, some vendors recognised the impact that these technologies would have on the OSS stack. With network operator business needs in mind, they identified the need for a new OSS stack that could take advantage of the enhancements brought by these technologies. These changes provide these experienced companies with new business opportunities for themselves. To address this opportunity, some have started to develop a new OSS structure that integrates Cloud, NFV and SDN technologies through APIs into a newly structured OSS stack. These technology changes have now enabled vendors to cross over into the OSS domain.

To many, this may seem to be a very complex and almost inconceivable change. It is the understanding of many people that changing an OSS stack takes many years. However, with the shift of focus on the OSS stack and some of these functionalities being taken up by other technologies such as Cloud, SDN and NFV through their orchestration layers, change no longer has to take a significant amount of time. This consideration needs to factor in the understanding, that much of the change will only involve a restructuring of the OSS stack and enhancing its focus, rather than the creation of a completely new development.

10.1 The changing focus for the OSS

OSS has consisted of four main components: orchestration, provisioning, assurance and fulfilment. These four components have had the role of identifying, creating and enabling configuration into the network and to identify how the service in operating when active in the network. As SDN, NFV and Cloud can now address the roles of identifying, creating and enabling configuration into the network, the OSS stack can now concentrate on how the service is operating. For this, SDN, NFV and Cloud orchestration systems need to report the state of the service to an OSS customer management and other SLA focused systems.

The following description gives a high-level overview as to how these systems can be structured and will function.

Service provisioning

Open northbound APIs in the SDN architecture provide the interconnection from systems such as the service catalogue and/or product portal (from the OSS and BSS layer) to the SDN Network and Policy controllers. These systems can trigger a template-based configuration that can be abstracted onto any end device, irrelevant of device type, command line required or version of software or hardware. To do this, the SDN Controller abstracts the appropriate configuration for the end node depending on the southbound protocol it uses. It is feasible to configure a wide range of diverse equipment – from core to end customer equipment. Included into the SDN Controller solution are applications that deliver resource management, resource control, and resource allocation. These resource applications are integrated into the SDN Controller for the automated creation of the appropriate templates long with allocation of customer and service specific resources. With the ability to hook to the service catalogue this permits for real-time changes to be applied, using features defined within the service catalogue to deliver the growing business requirement of customisation. As the SDN approach introduces a new methodology for service and infrastructure provisioning, the strategic OSS stacks now don't need to retain a focus on this function anymore for network and cloud infrastructures. This policy controlled configuration approach can now be considered for greater numbers of services such as VoIP, TV, etc.

Service assurance

This has two functions.

- The first is service management within a business logic context. This is further supported by Big Data analytics.

- The second function is to address service management of the customer within the network. Some SDN vendor solutions already include service management capabilities that can already identify basic service faults, and initiate fault resolution or work processes to address these issues.

Within an SDN Strategic Architecture, it is expected that the single focus of service assurance will be on delivering compliance to business logic - not carry out two very separate functions as it does in current network operator environments today. Analytics will be received from network elements and CPE, and this data can be analysed using Big Data to provide the information relating to how the service is operating. This ability to focus on ensuring the customer service is managed according to business rules enforces the priority of achieving revenue for the business while fostering customer satisfaction.

Now that analytics can be gathered by the protocol and/or API from the end device and intervening network nodes the OSS stack gains visibility of a new level of data for service and infrastructure management. The superiority of this enhanced management data – compared to the limited view SNMP or syslog have historically been able to provide to the orchestration, assurance and fulfilment systems - enables the OSS stack to generate a more sophisticated level of service management. This data can now be made available to Big Data systems for analysis and structuring, which liberates the strategic OSS stack development to focus on Big Data and its integration into the business and service management.

With SDNs focus on service fault identification and resolution, this permits for this the secondary function to move from the OSS environment and into a proactive NMS environment for customer service control.

Service fulfilment

SDN enables a policy-based configuration control, which permits for new levels of automation that have historically not been feasible. A fulfilment solution is still required within the OSS architecture. However the SDN policy-controlled environment requires that the data collated is dealt with by the fulfilment system in real-time. This requires the fulfilment system to interface to a Big Data analytics solution for accurate analysis of service fulfilment.

As different SDN Controllers support different capabilities, some which include topology, network inventory and resource management, the nature of the fulfilment system in an SDN environment depends on the features and functionalities supported within the SDN Controller solution that has been chosen. As fulfilment systems need to ensure monetisation of assets and to support and inventory of assets, this data should not reside within the fulfilment system, but in the Big Data store. This data is valuable to countless other aspects of the business, and collating it in a system from where it will need to be exported is a costly exercise. The LNMF network modelling function provides new capabilities to the fulfilment solution. It enables the forecasting of demand, analysis of future customer usage and efficient use and development of assets.

Service Orchestration

SDN, NFV and Cloud are architectures that incorporate and deliver their own orchestration layers. These have a focus on both infrastructure and service orchestration. Utilising the layered API approach, provided through SDN and by using policy and network controllers, these architectures integrate and deliver service orchestration at a technology architectural level. An example of architectural orchestration is where the service catalogue capability, interconnects using APIs to the SDN network and policy controllers to deliver flexible service control. The SDN systems then trigger customer products in real-time onto devices in the network.

Service Security

Security is key within all businesses. In an SDN Strategic Architecture, it should be considered that a security functional block should be included into the OSS layer to address the needs of the customer and infrastructure. The security function should be considered as being a primary function of the OSS, much the same as assurance or fulfilment. With an OSS security function the control and management of business logic, business policy and legal obligations can be incorporated at a service level into the overall architecture. This becomes feasible in newly commercially focused

OSS architecture, because it is at this point that business and technology logic come together, and where the data is available to make the required decisions to enforce change into the customers' services.

Such an expansion of the OSS infrastructure would ensure the security of the customer and the compliance of the network operators' business to the network operators' security policies and procedures.

SDN brings new capabilities and flexibility in solving many of the security concerns involved with the management of the consumers' services. Through the use of functions such as vCPE, and by having the ability to validate the compliancy to a configuration by using a policy- based configuration check, solutions can be created that validate whether the policy has been modified by the consumer or by a hacker, and confirm access rights to the content or the systems.

Now that the configuration is simpler to apply to the end device, a service security and privacy function can grow - for the protection of the end user from both themselves and from others. This opens up new revenue opportunities; it allows for network operators to differentiate themselves across the market and to have a layer that clearly focuses on ensuring compliancy to business logic, business policy and legal obligation.

OSS service control considerations

An SDN central controller enables applications to interconnect using APIs to respond to a direct query in real-time. With appropriate service and infrastructure applications in place, this can allow for the automation of fault and service problem analysis to enable automated corrective actions.

When architects are evaluating the required end-to-end architecture, they need to consider how relevant applications can be delivered. These developments can be driven through internal development, Open Source, or partner relationships with vendors. Key business decisions need to be made as to the importance of these systems, and how they will benefit future product development and differentiate the company within the market. If the concept for the applications being considered will significantly differentiate the future business products for the network operator, it may well be in the interest of the network operator to install an internal development and to retain sole control of the solution, much as the Web Platform companies do today. This will require internal development teams within network operators. Many network operators do wish to achieve the flexibility of the Web Platform companies. To reach this goal, restructuring of staff is insufficient – it will be required that the software development work is also done in-house. Web Platform companies do deliver their own software development on a large scale for their business-focused applications, whereas this is not something that many operator companies have done at the same scale.

Some SDN Controllers replace the need for the OSS systems to evaluate a multitude of different vendors' products, different command line structures, different generations of equipment, different code releases etc. This removes the need for considerable functionality and complexity from the OSS architecture, and requires the OSS architecture to interface to the SDN Controller via the API. OSS systems can receive data that identifies the state of the network, and modifying the end device configuration in order to update the service.

As most IT professionals are aware, the APIs used in SDN are not standardised. Some SDN Controllers use the REST API, with other APIs being considered as needs evolve. Through using open -APIs and SDN Controllers, a properly constructed query or reconfiguration instruction can now be abstracted down to the network through an open north-bound API from a management application or from a service catalogue portal.

OSS Service analytics considerations

To release the full potential of SDN, Big Data analysis and recording is required to be integrated into the overall OSS architecture. This will prove convenient timing, as many organisations have been investigating how best to utilise Big Data systems and how Big Data can be incorporated into their working environments for immediate and future benefit. Big Data records and Big Data analytics can be structured, so as to produces data analytics that can be sourced and used to enhance the management of both the customer service and the infrastructure.

The purpose of this architectural strategy is the development of an architecture that can create, implement and manage a very flexible service for the customer and for the network operator. Considering how the operator industry is shifting from offering a few products that serve millions towards product customisation, the OSS architecture is now about ensuring that a combination of components building blocks can be tailored to the consumer's personal preference, managed and reported on. These product building blocks need to be constructed in such a fashion as the consumer can individually select them. By having the option to select their own personal combination, millions of product combinations are created that may only be delivered to a few subscribers.

To achieve this, the architecture is structured using concepts taken from an Object Orientated approach. This allows for scaling and is expected to support the future evolutions of the architecture.

10.2 SDN/OSS integrated Architecture

The SDN communities desire to resolve the issues concerning the infrastructure and service management, needs to incorporate the lessons learnt by the OSS experts over the last two decades. For SDN to move forward, the IT silos need to be fully involved in the creation of the next stage of the end-to-end solution. To understand the context of the changes, much consideration needs to be given to the fact that networking solutions through SDN have solved many of the limitations that have constrained the ability of the IT systems to deliver real service management. Today, with exposed analytics from the network through APIs and with the significant work already done on Big Data, a new way can be considered.

When looking at these changes, it needs to be taken into account that many companies don't want SDN, NFV, Big Data and Cloud to change anything on their OSS stack - but at the same time they wish to realise the full flexibility of the working practices of many Web Platform companies. To get to this level of flexibility, it needs to be recognised that retaining a bottleneck, which was designed for a different approach to customer management, will restrict the capabilities of their business and the ability of the SDN Strategic Architecture to deliver. With networks and IT systems changing significantly and companies having their own specific goals, management and architects will need to be able to set out and explain to the teams the long-term strategy to achieve these goals.

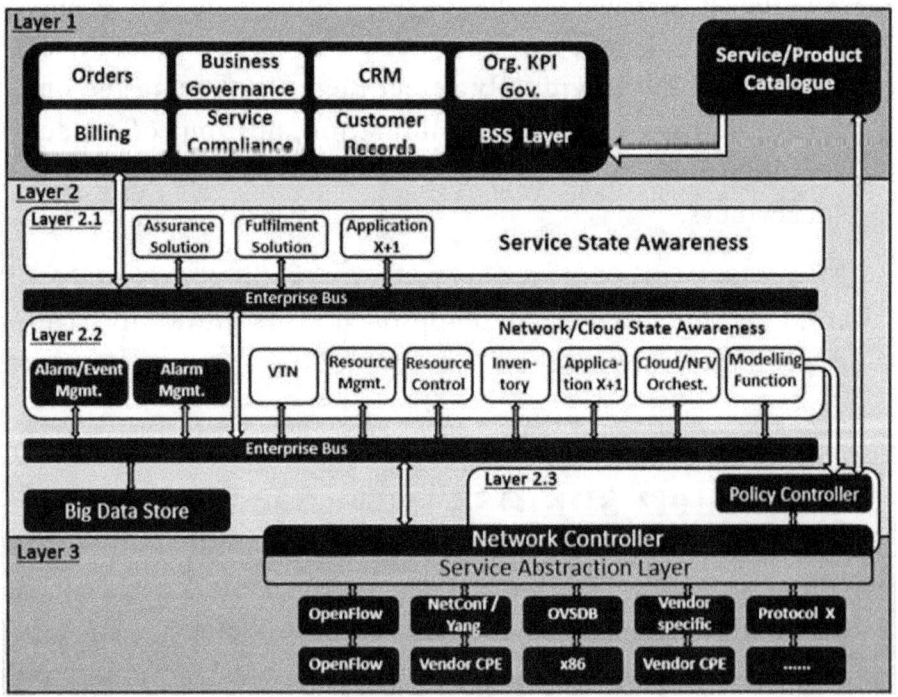

Figure 27: Overvew of integrated architectural control

The solution developed should not constrain the business and permit for the locking away of the data generated through open APIs and by the applications used for the management of the network operators environment. Data needs to be retained in a structure that is accessible by any part of the organisation and to not have it

154

locked away in a vendor application (be it an OSS or another application). As part of this overall OSS review, the following section highlights suggested thinking on how Big Data can be used to ensure that the data generated can be made accessible to all of the network operator systems. The aim is to generate an architecture that will be scalable to meet the demands of the information age.

The key purpose of this diagram is to highlight the need for a strategic end-to-end solutions architecture, combining the many technologies that are evolving in different areas of the network operators' environment. This architecture allows for multiple departments and roles within a network operators' organisation to operate in the much required cohesive fashion.

The diagram is not intended to describe all aspects of the end-to-end solution as different network operators' companies come with differing requirements. Instead this approach provides a high-level overview of what the SDN Strategic Architecture can create, within a long-term development strategy.

High Level end-to-end overview

The aim of the architecture is to achieve the delivery of a layered approach. The benefits that a layered approach gives are that it allows for the clear separation of the functions that utilise controlled exposure of data through APIs.

Using management applications and SDN components provide for the automation of service enablement, service status reporting, activation of proactive fault resolution (or informed notification to the engineer with a RCA), the automatic configuration onto the network, and the control of the flows which make up the service, in either real-time or near real-time.

The SDN Controller and its evolving tools are themselves automation tools that come with an end-to-end view on the network. Their purpose is to react to instruction sets and to automate depending on a situation. There is an excellent discussion as to be had as to whether these are in fact OSS tools. The question now to be asked is: Should the network controller go into the OSS, or should the OSS be positioned with the SDN Controller applications? By including the OSS function into SDN, the end-to-end business logic of service management and control becomes unified. The diagram assumes that the NFV network functions will be fully integrated into the SDN and Cloud orchestration functions through the policy control, and that their management will form the change control for the network requirements.

End-to-end Architecture: BSS Layer

This following diagram highlights some of the key components that make up the BSS layer. This section of the diagram is included only to explain to the reader that the BSS layer exists.

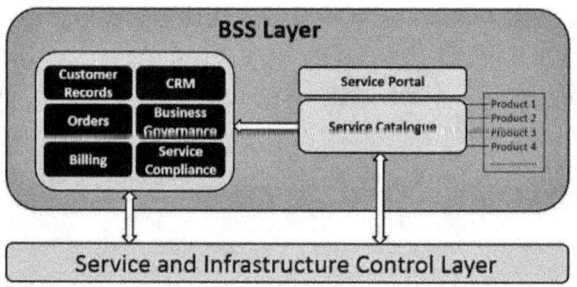

Figure 28: BSS Layer & Service Catalogue

With nothing of significance changing at this layer, and because there are many very well documented overviews available concerning the layout and structure of the BSS layer, almost no focus is given to this subject in this discussion. The BSS layer delivers the customer management and, through a product catalogue structure, allows for the automated instantiation of the service. This permits for the service to be broken down into its sub-components, and then to interface through an API to the policy control to allow for the creation of the service at the lower levels. These details are explained in the relevant lower layers. The service is then programmed into the network control layer via the service and infrastructure layer. The additional data available permits for a greater level of granularity of service control, management and reporting.

156

E2E - Service & Infrastructure Control Layer

This layer represents the interfacing of the SDN Controller, the historic NMS, the OSS and Big Data along with some new functions such as the Live Network Modelling function.

This occurs via the northbound interface. The purpose of this layer is the abstraction of the business rules and the instantiation of these business rules towards the network control layer. Data can be received that reports the success or failures in the service and the instantiation of the commands to reset the rules by which the service is controlled.

This layer has the function of identifying the state of the service and the infrastructure through collecting and processing the management data from the network elements using APIs. The goal is computing a resolution to problems that were identified for both the service and the infrastructure. Layering puts a structure in place to enable governance, structuring and utilisation of service and infrastructure data.

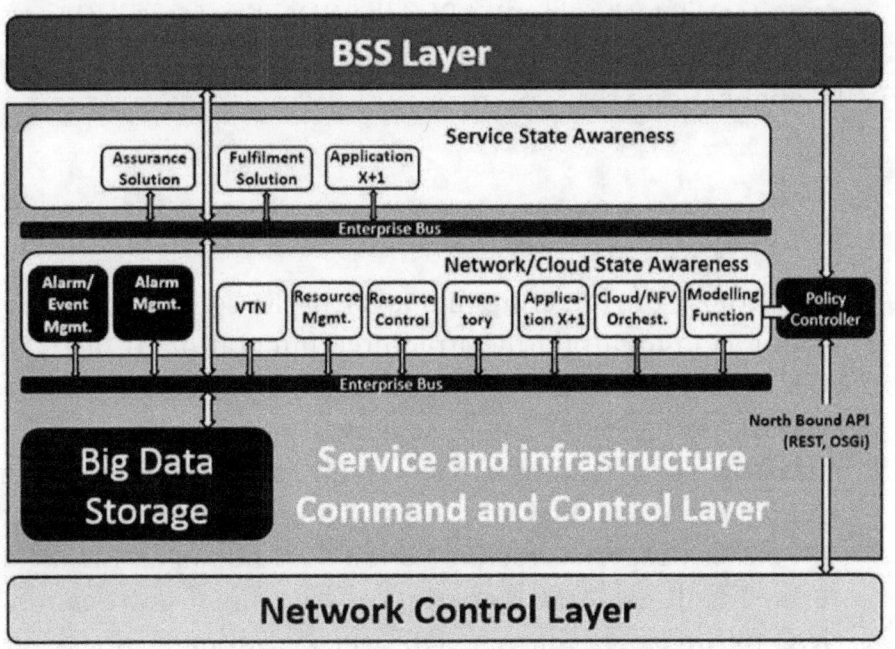

Figure 29: Service and Infrastructure Control Layer

157

For this layer 2, the architecture is broken into three sub-layers, as all these solutions interconnect with each other to ensure service and infrastructure control. These sub-layers are described as follows:

Sub-Layer 2.1 - Service State Awareness

This layer deals with the OSS functions. It concentrates on the business-focused service awareness, and is expected to operate in near real-time. It utilises fulfilment and assurance functionalities similar to those that are available today. It utilises enhanced analytics capabilities to create a solution that can deliver service state compliance to the business SLA. The IT Systems, enterprise bus architecture is expected to be used to interface all the systems of layer 2.1, 2.2 and 2.3, therefore permitting for any-to-any connectivity between applications and Big Data. Where required and to ensure the forwarding of the relevant data, APIs are used to interface the necessary systems together to make sure the relevant data can be communicated to the appropriate systems. With SDN using (an IT technology) APIs, the SDN Strategic Architecture aims allows for a unified integration interface to integrate networks and IT applications.

Sub-Layer 2.2 – Network/Cloud State awareness

The purpose of reflecting this layer in the service management section is to highlight the structured integration of the NMS and the OSS as part of the end-to-end strategic architecture. The layer 2.2 network management components will notify service and infrastructure affecting impacts to the layer 2.1 OSS systems. When such an event occurs, layer 2.2 systems will analyse the fault and will either trigger corrective action based upon an identified pre-defined fault use-case scenario, or will trigger notification to the NOC for investigation. The OSS stack with its business service SLA focus is also expected to identify when the total business SLA is being impacted. E.G. A series of service faults have occurred over the monthly billing period to a customer. Proactive service fulfilment analysis has identified that further impacts to a service would result in e.g. loss of revenue to the network operator, an evolved OSS system would trigger a request to the Layer 2.1 systems to reprogram the network, so that all relevant service traffic had the highest priority and would therefore make sure that the service had a reduced opportunity to be impacted. As end-to-end SLA information can be included now, service management can become proactive, without having to continue using the highly costly approach of network overbuild to ensure the adherence to SLAs. The Big Data generated about common faults can be used by the operational teams to identify areas of the network that need to be attended to, and to develop use cases for proactive fault management and resolution.

Please refer to the chapter "SDN: Service and Infrastructure Architectures" for the discussion about this layer. It is also near real-time, but will have components that will evolve to real-time capabilities as technology evolution progresses. How this architectural approach is incorporated into the end-to-end architecture, is addressed in chapter 11, End-to-end description: SDN Strategic Architecture.

Sub-Layer 2.3: Policy and Intelligent Control

This layer is isolated from other near real-time components due to its real-time nature. It is the single point of initiating change into the network. This removes the complexity of multiple systems triggering changes into the network and allowing the controller to become out of sync with the rest of the network. For a full description of this sub-layer and the different solutions architectures please refer to chapter 7, An outline of the functional components. How this solution is incorporated into the end-to-end architecture is addressed in Chapter 10, SDN: Network-integrated OSS Architectures.

Big Data

This structure is isolated from the rest of Layer 2, as it interfaces too many domains and supports the service awareness of the architectural approach. It is used to collate the data gathered from the component devices and to provide data to the various components. The intention of this architecture is that the data gathered is continually processed by structured analytics engines, to create data structures that can be used by the applications that manage the network and the service.

These function-based management applications can therefore utilise the Big Data system to extract and to analyse exactly the data that their function requires. In a next step, this data is pushed back into the Big Data structure, allowing other applications to make a call to this processed information. Ideally, this would permit the Big Data store to serve as the source of all information and to be the destination to the storage of all analytics. This creates a central repository for data, with the relevant data being stored in appropriate data structures and from where the data is requested by the applications.

Accessing the Analytics

Historically, much data was destroyed due to aggregation or, in the case of routers, could not be obtained due to the lack of detailed APIs on the platforms.

Few network operator businesses have the same business models or business goals. These differences lead to the creation of different data structures to resolve the needs and the issues of the organisation. For this reason, the following diagram is given only as an example. It suggests a model that could be used for the structuring of the data gathered from the network using the SDN API capability.

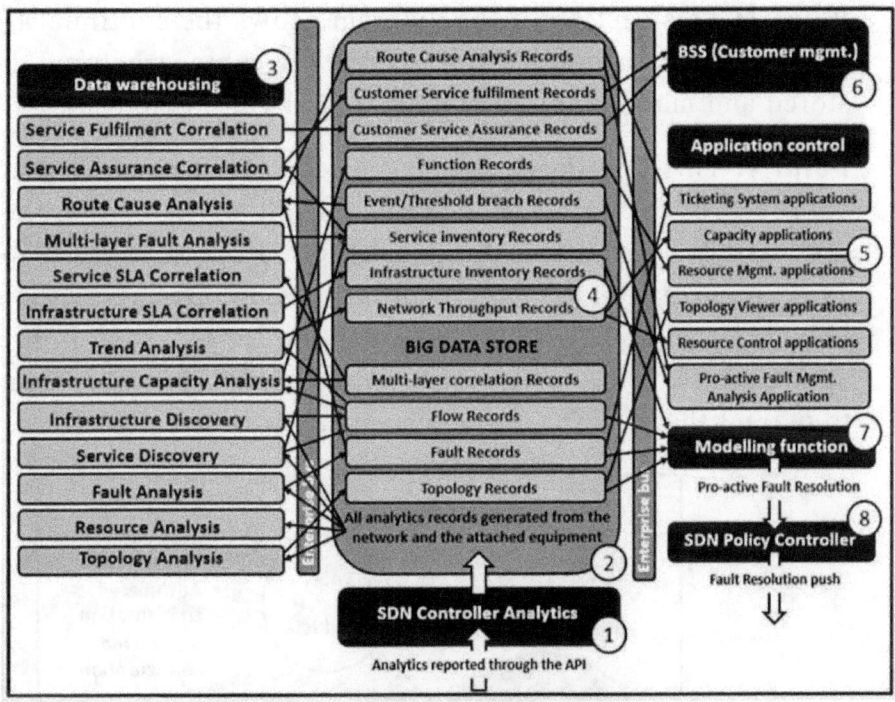

Figure 30: Big Data gathering and structuring

Note: The diagram above only shows an SDN Controller gathering analytics; it can be assumed that there are many other sources of Big Data.

This diagram highlights why an architected structure is crucial for dealing with the generated data – data about the network infrastructure and the service state. It casts a light on the need for solid data structures that still can be modified, and on the importance for a company to retain complete control of its data and not to have it residing permanently in proprietary applications.

The following is a point description of how Big Data could be gathered and structured.

Point 1: This shows the collection of all data through the SDN APIs (or other inputs) in a raw format. Data from other sources such as Cloud and NFV infrastructure or customer end devices can still be incorporated.

Point 2: The reference points indicate the general Big Data storage

Point 3: This section of the diagram shows the continual Big Data analytics structuring of data, to create useable structured data that is stored and made ready for relevant applications to use.

Point 4: This point reflects the creation of the specific granular data records for applications to use, while ensuring that these records can be made available to all other applications to support evolving demands for data records. This avoids having the most valuable commodity of any organisation locked away in proprietary applications, where proprietary vendor development is then required to extract the data for other systems to utilise.

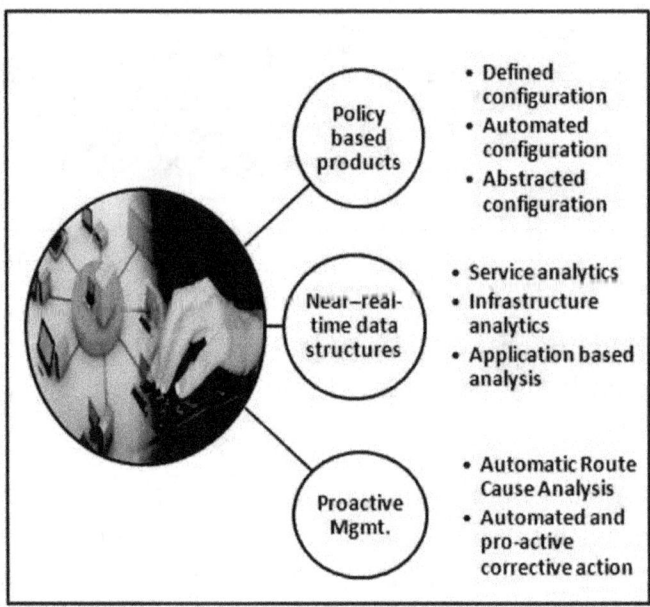

Figure 31: Data Driven Operational Solutions

Point 5: This reference point indicates the applications that utilise the data to create solutions and to inform the network NOC teams of the network and service state. The data generated supplies information to applications that can now process this data with the intention of creating proactive fault resolution.

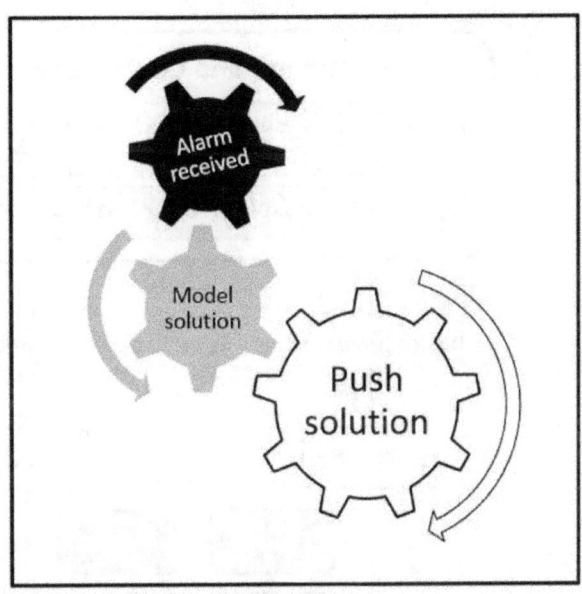

Figure 32: Automating fault resolution

Point 6: This points references to the creation of real-time or near real-time service records for use by the OSS/BSS applications to ensure accurate management of the customers' services and control of security.

Point 7: This highlights the inclusion of the Live Network Modelling function as an application that is integrated into the SDN Controller for analytics purposes. Data collected and created by this analytics tool could also be pushed into the Big Data storage for reuse by other applications. The LNMF can be enabled to analyse and identify solutions for problems which are occurring on an end-to-end network level. This specific function could be enhanced to use the inputs from the Big Data storage. This multi-layer correlation capability can identify resolutions for faults in the network, e.g. loss of fibre path, loss of router, loss of routes, loss of flows, trending towards congestion on a link, etc. These faults will be viewable, and when these events are identified and computed, proactive fault resolution can be triggered to generate corrective actions, such as initiation of a new optical path, signalling an alternative path on a segment routed flow, to name just a few.

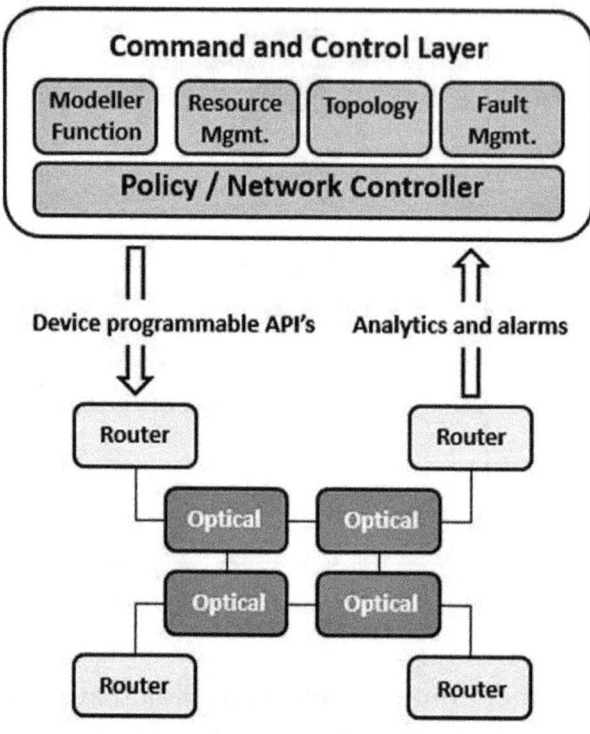

Figure 33: Integrating fault resolution

Point 8: The actions can be triggered through the Policy Control interface, or a ticket can be triggered to the NOC team with a Route Cause Analysis for actions to be carried out, depending on the available capabilities in the network.

E2E Architecture: Network Control Layer

For a full description of this sub-layer, please refer to Chapter 7- An outline of the functional components, for a description of the new SDN control architectures. How this solution is incorporated into the end-to-end architecture is addressed in Chapter 10, SDN: Network-integrated OSS Architectures.

The Network Control layer focuses on the automated provisioning of network and Cloud infrastructures according to the needs of the service. Within the SDN Strategic Architecture, the OSS layer interfaces to the API supplied from the SDN, NFV and Cloud orchestration layers. The OSS is no longer goes required to program devices deeper in the network. The type and structure of the API should be discussed with vendors and the CTO groups across organisations.

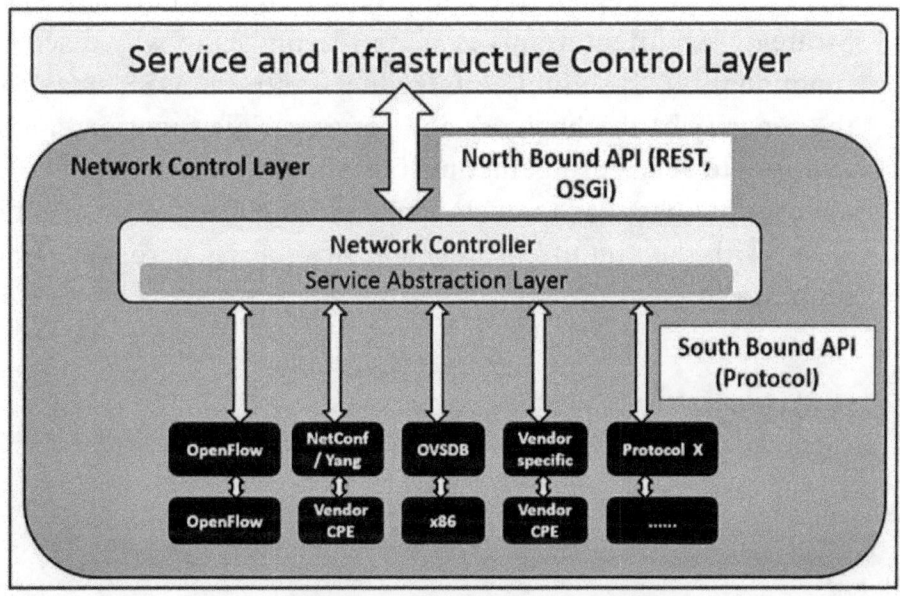

Figure 34: Network Control Layer

Summary

The SDN Strategic Architecture describes a strategic approach for infrastructure and service management. It targets the enabling of service management and control by focussing on delivering according to the business rules defined in the SLA. To deliver this capability and to control OPEX, the OSS and NMS are interconnected and are enhanced to intelligently permit them to communicate with the infrastructure architecture. This strategic architecture has been defined and structured to enable the creation of a proactive network maintenance and management solution. Through having SDN as the linking technology and by incorporating Big Data, the OSS stack can use the SDN Controller to trigger corrective action using real-time policy-based configuration.

SDN, along with NFV and Cloud, is a strategic direction to address a significant part of the service and infrastructure provisioning, along with much of the technology orchestration of the service.

The OSS stack can now focus on two key elements: business security and business service delivery. Commercial service fulfilment and assurance are well defined in the OSS stack, but the enhancement of security for the business by the inclusion of business logic, legislative and product logic is not. The SDN Strategic Architecture gives some reasons as to why a security functional block should be introduced into the OSS stack to control the security of the business and service. This function can be used to validate services against business logic, and to trigger appropriate changes to customers' products and infrastructure using SDN, with the aim of enforcement of security across the end-to-end organisation and the services sold.

11. End-to-end: SDN Strategic Architecture

The previous three chapters (chapters 8, 9 and 10) examined and discussed the three key focus areas of the SDN Strategic Architecture. This chapter now explains further and brings together the thinking behind the SDN Strategic Architecture and unifies the introduced functionalities into a flow-through layered architecture.

The aim of this proposed architecture is to identify a flexible framework, from which to evolve a next generation network operator architecture based upon the changes introduced through SDN. The reason for structuring the book in this fashion is to clarify the purpose and goals of each area and to identify some of their components.

As most network operators have different business goals and aims, only a subset of applications and components are considered, because as always, the requirements of the individual network operator are based on their individual business goals. In addition SDN and NFV orchestration functions are not addressed in detail as these components are well documented and called via APIs that specify the business goals and aims of the network operator.

Migration principles of the SDN Strategic Architecture

The aim of the SDN Strategic Architecture is to target the creation of an object-orientated, end-to-end, agile solutions architecture. The purpose of this approach is to enable businesses to gain a flexible architecture that evolves towards a self-managing, self-maintaining and self-expanding network within the focus of achieving a service-aware network and IT environment.

The reason of targeting this initial architecture, rather than defining a final architecture, is that the SDN Strategic Architecture delivers an adaptable framework. This is based on building blocks (objects and classes) that are capable of absorbing and incorporating additional functionalities (through APIs) to meet the needs of new and evolving business requirements.

Many of the concepts used in the SDN Strategic Architecture come from the IT environments and are therefore tried and tested technologies, even though they are not common in the network environment. The incorporation of tried and tested technologies and the combining of lessons learned from across the technology industry, enables the creation of a long-term evolutionary business-focused architecture that can be structured to the individual goals of a specific business.

The architecture has been defined with the expectation that network operators will continue to modify their networks and IT systems. This continual modification occurs because of re-scaling due to growth, the adding of new functionalities, the enabling support for new products, etc.

This provides the opportunity to strategically redirect the network operators' architecture and to limit major disruption or to require whole scale replacement. This permits for a strategic approach to be used when realigning the infrastructure and services, to meet the changing business needs.

The goal of the SDN Strategic Architecture building block principles is that a change can be achieved using a migration strategy, and by strategic architectural compliance and governance during new projects and upgrades.

To achieve this, it is required that new and evolving architectures and designs are continually reviewed to make sure that the aims of the strategy are progressing. This will ensure that all the relevant departments understand the aims and goals of the strategy, and that the strategy is explained to those who are expected to deliver the solutions to the business.

Layers of the SDN Strategic Architecture

The following diagram shows the end-to-end layered SDN Strategic Architectural approach. This approach reflects the separation of key functional systems that are used within a network operators' network to deliver services to customers.

For simplicity of the diagram, the full Cloud and NFV orchestration systems components are not detailed as these are self-contained within their own structure. These self-orchestrated components are expected to interface through APIs into the infrastructure management layer (Network/Cloud State Awareness) layer.

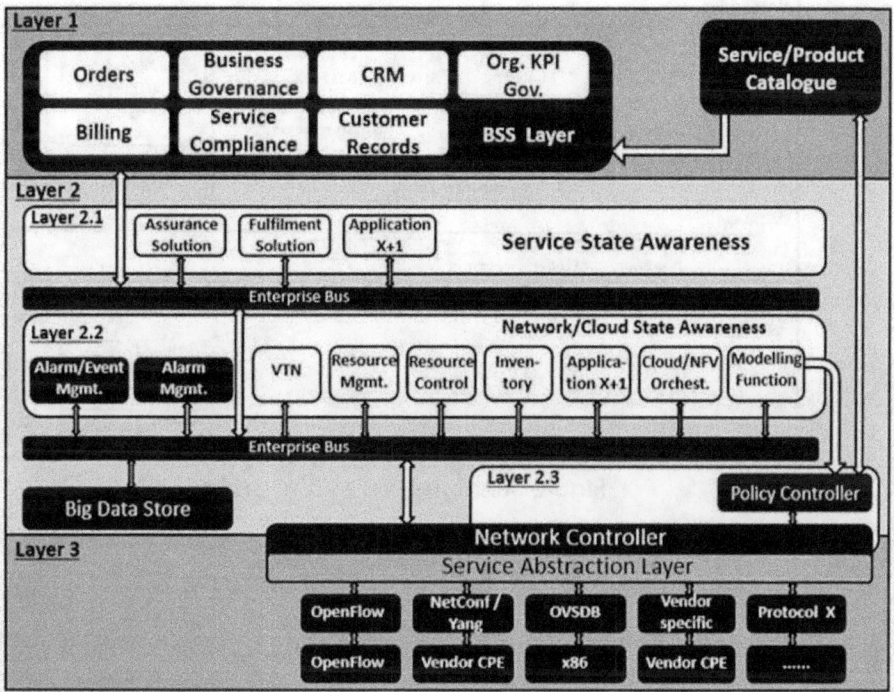

Figure 35: Identifying the Layers of the SDN Strategic Architecture

Call flow through the SDN Strategic Architecture

This flow diagram represents the setup of automation through the architecture. The identified flows represent how new functional developments can define an architectural structure that enables new levels of automation.

Additionally, the flows show the population of customer-relevant data and SLA information into the relevant layers of the

169

architecture to ensure provisioning, monitoring, management, and SLA assurance within an end-to-end automated architecture.

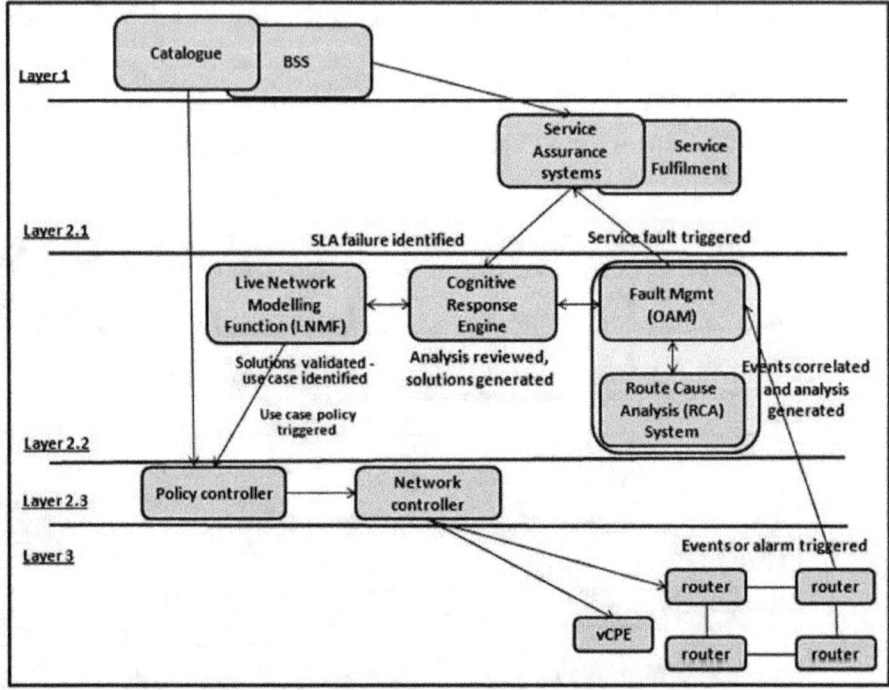

Figure 36: High-level end-to-end control flow

11.1 Layer 1: Service Enabler (BSS) Layer

An overview of some of the components within layer 1:

- Customer portal interfaced to the service/product catalogue

- Customer records

- Customer personal preference records

- The usual BSS system functionality

- Micro-payments system

Layer 1: functionality description

When compared to existing BSS architectures and systems, little changes because of SDN Strategic Architecture at this layer. Only the key functional systems of importance to the description of this

170

layer are listed. These are common across most network operator environments.

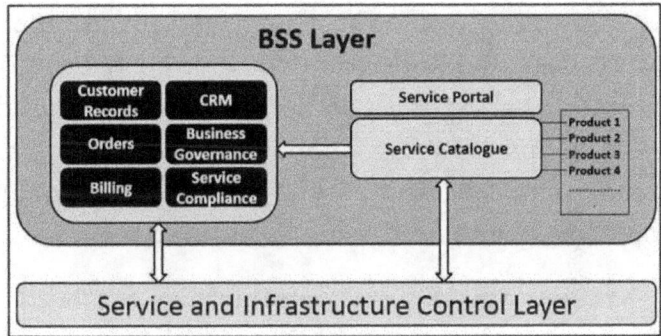

Figure 37: BSS Layer & Service Catalogue

Many network operators already have service/product catalogue functionality in their BSS architecture. This functional system can be used by network operators to help realise the benefits of an SDN Strategic Architecture – by enabling new revenue streams, and by enhancing new product opportunities. The service catalogue refers to the service descriptions that are viewable by the consumer and which can be utilised or purchased.

This provides the ability for enhanced automation provided by SDN using policy and network controllers from the service/product catalogue function. This enables the ability to trigger real-time service activation on a granular level into the network infrastructure for customer service configuration.

This ability to trigger automatic changes through the policy and network controllers allows for an immediate change to be triggered from the operations teams, the customer, or from the BSS systems as an update to an entire product suite.

An important enhancement that this approach delivers is that it provides the capability to deliver real-time customisation of the consumers' experience, while, at the same time, permitting for the information to be triggered into the service management layers for SLA fulfilment and assurance.

Service customisation towards the customer may require micro-billing functionalities within the BSS solution. As many operators already have both micro-billing functionality and service/product catalogue platforms enabled, these systems can continue to be used and expanded on.

Any changes required to these applications are therefore expected to be feature-based and not architectural; however modifications may be required to APIs. With an appropriately featured customer portal, SDN, Cloud and NFV can support customisation, as customer preferences can be gathered via the service/product portals from the customer and configured into the service.

As the systems and APIs within the SDN Strategic Architecture can now be structured to report the required analytics from the network, the OSS assurance and fulfilment systems can access the data relevant for the management of the customer.

With the correct applications, this permits for proactive fault analysis to be run against the network and services to identify existing or developing issues. In a next step, this information ensures that the services are delivered in line with the customer product and preferences expectations.

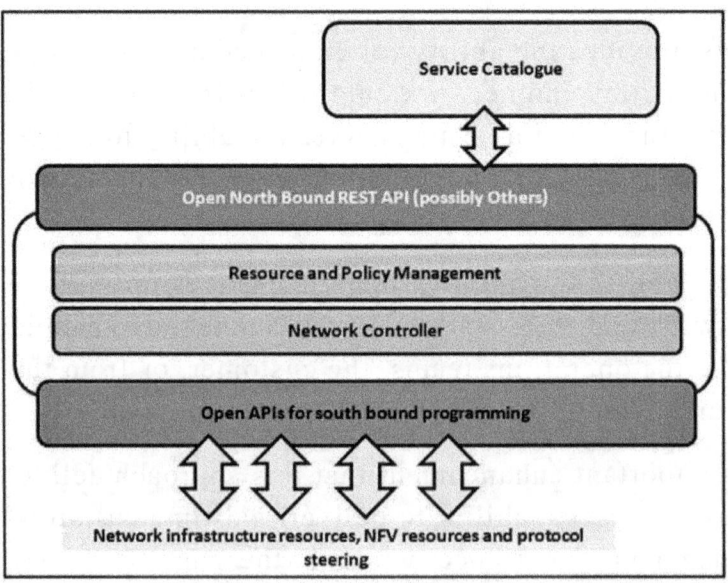

Figure 38: API interfaced dynamic service management

With the inclusion of the new functional systems at the lower layers of the SDN Strategic Architecture, the product definition and its associated SLA can now become the terms and conditions for the management of the IP flow-based service while it is active on the network.

At the point of service creation, the relevant product specification and customer preferences information will be required to be pushed southbound in a secure manner. This can be achieved through an API to Layer 2.3 for the policy control of the customer vCPE instance. The local customer policy controller within the vCPE reacts to analytics that indicate a breach in the customer service SLA. This ensures the service is managed according to the SLA, and, for the service SLA adherence confirmation, information is fed back via an API to Layer 2.1 (OSS) as well as to the BSS stack for billing and registering of SLA adherence for the service instance. The terms 'service and product catalogues' are standard industry terms and therefore industry definitions exist for these components.

11.2 Layer 2: Service and infrastructure Control Layer

Layer two integrates the historic OSS and NMS systems for the purpose of service and infrastructure control and management with near real-time and real-time service control capabilities.

This layer is structured into sub-layers and introduces new functional systems to drive automation and proactive control of both the infrastructure and the service. The goal is to move the network operator environment from its current state of reactive management and off-line provisioning to real-time provisioning with proactive management and service control. To achieve service management and control, OSS and NMS systems, which in the past have been vaguely connected or have operated in isolation, are now restructured to allow them to interoperate using bi-directional APIs.

APIs within the SDN and NFV architecture permits business logic to be communicated through policy to the controller using REST or other evolving APIs. The SDN controller abstracts the logic and applies it on a relevant southbound API protocol. This provides the appropriate instruction set to be applied to a device or group of devices, irrelevant of command line or vendors' equipment.

New application suites are required to be included into the SDN Strategic Architecture. These will provide for intelligent decisions to be triggered to the Policy/SDN controller, based on analytics analysis by the function-defined applications.

The diagram highlights the purpose of the functional systems that the SDN Strategic Architecture proposes be introduced at this layer. This identifies existing and new functional applications, which, when considered together, enhance the business control of the service and infrastructure for the network operator.

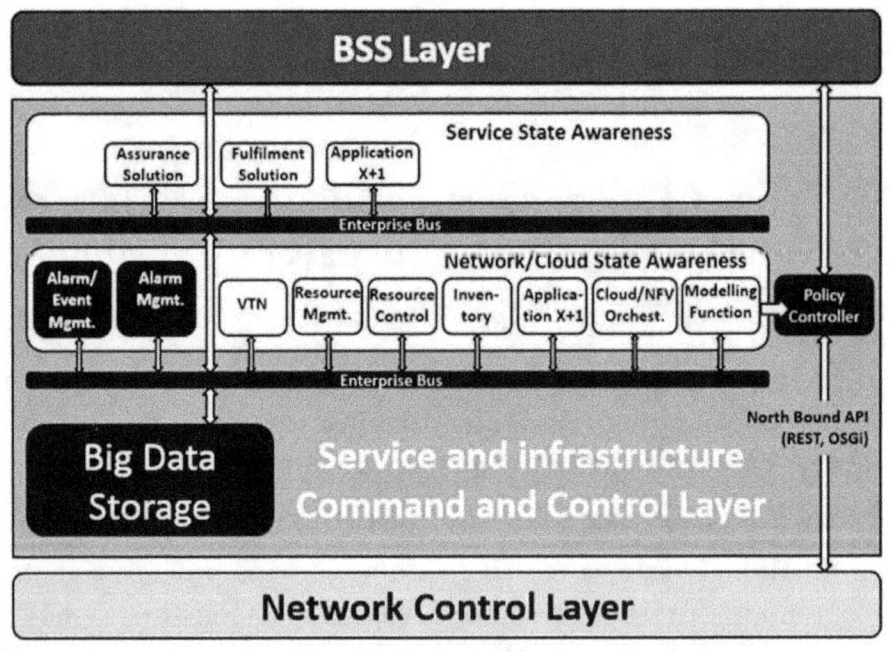

Figure 39: Layer 2- Service and infrastructure C&C function

With statistics collection from end devices being possible through APIs, the SDN Strategic Architecture enables an open approach to be used to influence services. Layering of the structure permits APIs to communicate through the enterprise bus, allowing for single points of control and decision-making. This enables controls as to which system can enable change into the live network, therefore preventing the management systems from becoming out of sync with what is enabled on the network.

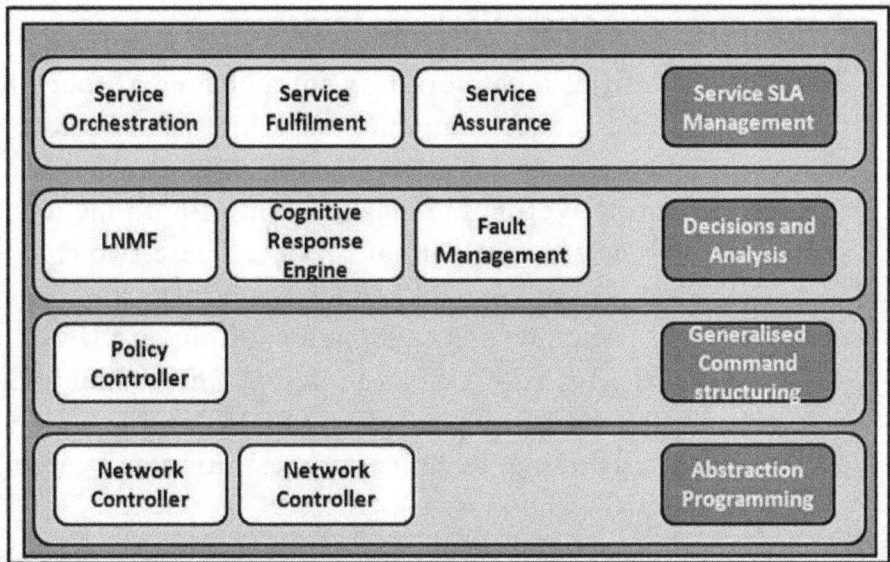

Figure 40: High-level layer 2 functions

The following sections describe the 3 sub-layers within layer 2 and the functionalities they contain.

Layer 2.1 - Near Real Time OSS Service Mgmt.

An overview of some of the components within layer 2.1:

- Cloud and NFV Orchestration

- Big data Analytics

- Big Data Store

- Service business fulfilment

- Service business assurance

- Analytics capabilities – e.g. customer analytics for targeting of new services, customer analytics to identify the value a customer brings to a company

Layer 2.1: functionality description

This layer identifies an opportunity for a change of focus for the OSS systems and suggests an approach which could create a functionality change at this layer. Historically, the OSS was responsible for delivering the management of both the technology infrastructure and the commercial service. These two capabilities were delivered through the OSS functions of provisioning, orchestration, assurance and fulfilment. Within an SDN Strategic Architecture, the focus of OSS is to deliver the commercial service management for the customer, leaving SDN, NFV and Cloud to deliver the provisioning, technical orchestration and gathering of analytics about the end systems.

SDN enables this by its use of policy-based provisioning, and by having the technology orchestration delivered through the SDN, NFV and cloud orchestration stacks.

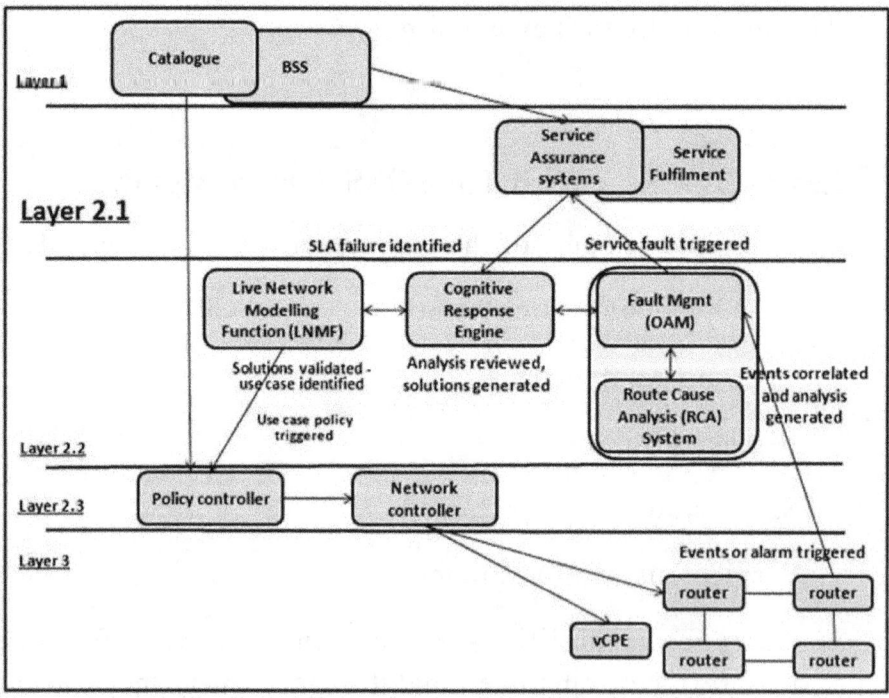

Figure 41: SDN Strategic Architecture - Sub-layer 2.1

Historically, IP flow-based service management and control has been performed by managing the end CPE or network elements. This is because hardware devices have not exposed sufficient

176

analytics for the individual per service-based IP flows to be fully identifiable within a network infrastructure. This approach is no longer sufficient in a world in which so many services now reside on the IP flow level, and in which operators have consolidated many of their historical service-separated infrastructures onto a single IP backbone. Network operator consolidation of network infrastructures has reduced operator costs, but simultaneously has reduced the network operators' visibility of IP flow-based services. Many of the services running over the network have different delay, jitter, packet loss, and other characteristics.

With consolidated networks having greater security exposure, and using aggregated forwarding rules, the IP flow-based service has now all but disappeared into the bulk infrastructure. It is therefore difficult to identify, fault-diagnose and to control without a new approach. With APIs and the external controller, the device will provide greater data for service analytics for control of the service.

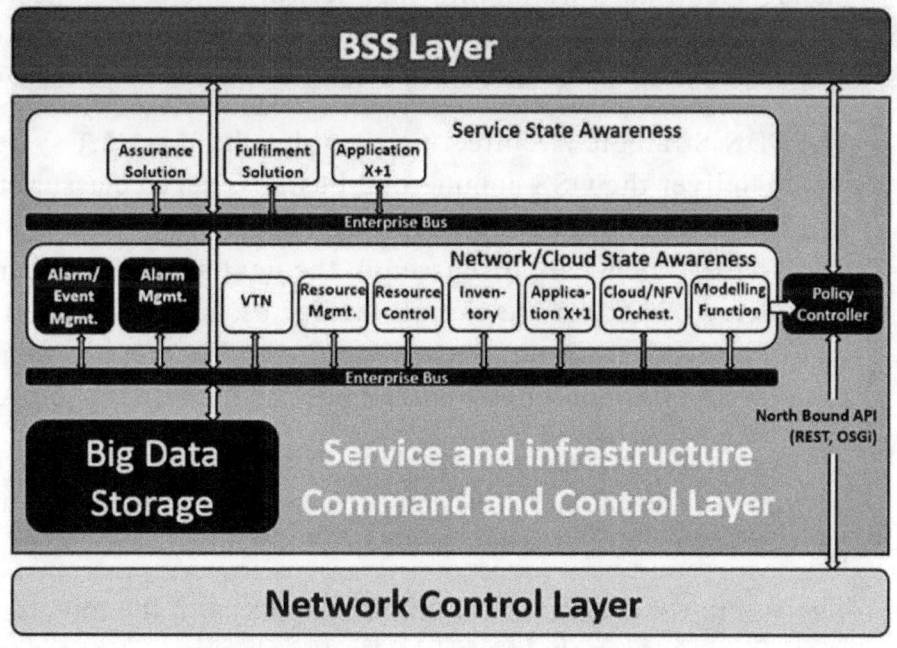

Figure 42: Service and Infrastructure control layer

SDN delivers the ability to change the responsibilities for this layer. Through policy-controlled SDN template-based provisioning, the historical end device and service activation capability is no longer required at the OSS layer. This can now be delivered through the

SDN Network/Policy controllers and abstraction functions. In addition, orchestration is further enhanced through the SDN, NFV and Cloud orchestration systems within their architectures.

The SDN Strategic Architecture requirements for OSS now have the singular focus on the commercial business service management. This is a business-critical solution, as these solutions provide assurance and fulfilment for the customers' SLAs.

Business Service assurance

Service assurance systems use policies and processes identified by the network operator to enable SLA monitoring and trouble ticket management. These functional systems are used in conjunction with policies and procedures, and these include fault and event management, Quality of Service management, performance management, network monitoring, network and service testing, network traffic management, and others. The aim is to identify, diagnose and resolve the issues affecting the customer service.

The SDN Strategic Architecture provides the layer 2.1 systems, which deliver the OSS commercial business service assurance and fulfilment functional system with new functional systems at Layer 2.2. These systems are discussed in the next section and can be considered as part of the overall service assurance function. In addition, Big Data is structured into the architecture to fully utilise the new level of analytics that can now be gathered using APIs.

Through SDN management, the analysis and control of the data generated from the network and the use of intelligent systems permits for the automation of much of the network-focused analysis.

The assurance solution is required to work with the new functions at layer 2.2. These include the LNMF, the Cognitive Response engine and root cause analysis capabilities. These applications bring a new level of intelligent decision making into network infrastructure and service management. They also provide for service assurance through their near real-time capabilities, and are able to identify and model the cause of the event.

These combined solutions now provide for automated service management. For example: The assurance system gathers data on the total number of impacts to a service over the period of one month. When commercial service impacts reach a predetermined threshold, the service assurance system can affect a change to the network to raise the service priority, therefore reducing the chance of it failing to meet its monthly SLA targets.

It can be argued that the OSS assurance function should be expanded to include the new functional components that are under development/consideration and as these applications will drive automation of the network and service management. The tools, when integrated, support and deliver both service and infrastructure management. It is therefore suggested that the assurance function concentrates on developing a full set of functionalities that will ensure full commercial service management and control.

The splitting of this major historical function therefore takes into consideration the major change that SDN brings. This casts the spotlight on service management by management of the IP flows. These IP flows are the streams of IP packets that travel through the consolidated network infrastructure. It is these flows and their adherence to the service characteristics of delay, loss, jitter, delay, etc. that make up the service and which, when transported within a set of boundaries, deliver the user experience. Focussing on managing the end-to-end IP flows that make up the service, rather than managing the infrastructure as a way of emulating service management, moves customer service SLA management into the centre of attention.

Business Service Fulfilment

Service fulfilment is the processes involved in assembling all the components of the service, and enabling the features of the service to be made available to the customer. Key to achieving this, is data being available to all required systems, automation of the configuration of the full service, and understanding the availability of resources.

This can be separated into two separate work classifications: technical fulfilment and business fulfilment, which equate to the fulfilment of customer service.

As with assurance, the future evolution for service fulfilment should concentrate on the commercial/business aspects of service delivery. The focus for the service fulfilment will involve service billing and management activation, service order receipt and entry, decomposition of the order into its constituted components, the workflow tracking of all components of the order, and the resolution of any service features that can't be delivered during the initial phase, whether it is due to logistics or the lack of capabilities on the network.

The technology aspects of service fulfilment, which should be addressed in Layer 2.2 through the SDN management functions, will include network configuration, network capacity management, and network inventory management control and allocation. The SDN template-based policy-controlled configuration allows for the delivery of the end device configuration based on generalised instruction sets that are generated from the BSS service and product catalogue functions. These will be connected via an API which provides the mechanisms to automate the fulfilment of the service technical delivery to the customer. This split in the historic OSS functions allows for a clear demarcation of responsibilities and deliverables, ensuring both a clear focus on the technical level through SDN and a clear focus on the business deliverables to the customer via the services management by the assurance and fulfilment functions at layer 2.1.

Layer 2.2 – SDN: Near-real-time Command and Control

An overview of some of the components in Layer 2.2:

- Command and Control

- Fault Management

- Ticketing system

- Root Cause Analysis system

- Resource Management system

- Resource Monitoring system

- Cognitive Response Engine

Layer 2.2 Architectural Integrity

It is preferred that the following functions are developed as standalone functional components. If they are developed as part of an integrated vendor management solution within the fault management or LNMF functions, the architectural structure has the potential to be compromised. It is the role of the architects and designers to ensure this doesn't take place during the strategic evolution.

If it does happen, it creates the opportunity for multiple vendors' products and solutions existing in the same environment with duplicated functionalities. This is a recipe for operational headaches when structuring all the data for a multitude of systems within the end-to-end system management solution.

As these solutions will become part of the usual feature race between vendors, it is crucial that the network operator retains control of the strategic architecture for these C&C functions.

Layer 2.2: Functionality Description

Network and service management are no longer an afterthought of design and within NFV, Cloud and SDN technologies and the SDN Strategic Architecture. Their focus on management and control is central to how they have been created. With this in mind, the SDN strategic architecture aims to deliver:

- A flexible and customisable structured solution that resolves the business requirements and business strategy

- The technology functions and a technology strategy that can meet the business goals

This sub-layer operates in near real-time, as it requires data to be processed to produce analytics. These analytics are then communicated between the functional systems at layer 2.2, and outputs provide triggers to layer 2.3 systems for the activation of solutions into the network.

The structure and capabilities of these new functional systems have been identified to enable call flow capabilities that permit the shift from reactive to proactive network and service management. APIs, bi-directional communications channels between the functional systems, Big Data and the network elements provide the live data that can be processed for decision-making in near real-time.

Additionally, APIs are used to communicate instruction information to Layer 2.1 and Layer 1 business-focused orchestration, assurance and fulfilment systems. The data is communicated to confirm the customer service state. This ensures there is a trackable and traceable record of the customer's service operating within predefined SLA boundaries.

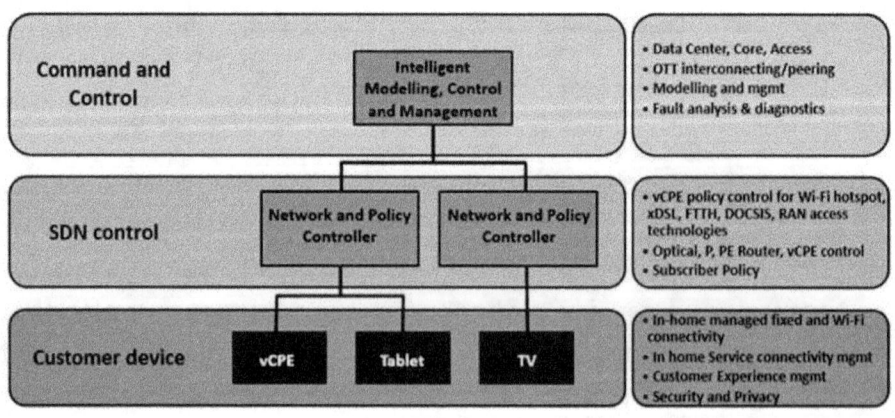

- SDN delivers the integration of network infrastructure control and infrastructure mgmt.

- SDN permits end-to-end control of the service across multiple historical silo's

Figure 43: Service affecting Command and Control

Sub-layer 2.2 delivers much of the Command and Control functionality with a specific focus of driving the delivering of near real-time control and management. Some of these functional systems have had early version of the software released into the market.

These systems still have reduced functionality and do not yet achieve all the targeted functionalities described in this strategic architecture. For the end goal features to be fully realised for use by the network operators, it will require that architects and designers drive the development of such solutions and achieve the functionality that their organisations' business model needs.

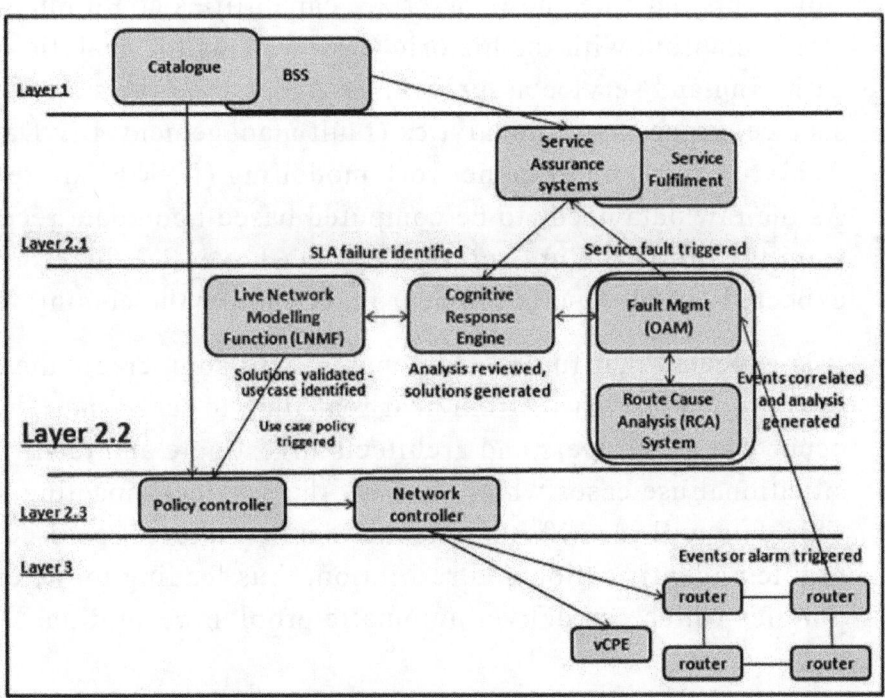

Figure 44: SDN Strategic Architecture - Layer 2.2 functions

This can be achieved by partnering with vendors and sharing requirements, by initiating an Open Source group to get industry-wide focus on these concepts, or by driving internal development. The methodology selected by the network operator will depend on various local factors such as cost, time for implementation, management commitment, etc. The creation of such technologies can be considered as a method to gain a very significant advantage over the competitors and therefore the method chosen to develop such solutions is linked to the strategy of the business.

Companies who have a long-term business vision may well choose to initiate internal development as they wish to ensure that they retain singular ownership of software for market advantage.

This does not stop network operators from also releasing older versions of software into the market and for the network operator to function as software houses for extra revenue generation through the sale of such software applications.

The diagram above reflects an example management call control flow. This indicates how Layer 2.2 capabilities could integrate the NMS function, with the historic OSS systems for analytics gathering and service assurance. To understand the state of the service, processing of analytics (fault management, Big Data Analytics, etc.) and live network modelling (LNMF) are required. As the raw data needs to be computed based upon data received from the infrastructure and service, sub-Layer 2.2 therefore is expected to only operate in near real-time for the coming few years.

It is expected that future development will soon create the ability to evolve some use cases from near real-time to real-time. This will occur when engineers and architects investigate and identify the situational use cases which identify the service-impacting scenarios. This data will identify the relevant data structures required for problem identification and resolution, thus leading to the creation of policies which can deliver automatic problem resolution.

Command and Control

Across the industry, everyone is being asked to do more with less, and the same goes for the circuits that are being installed. Many of the problems on the Internet can be addressed by speed and buffering, and for many years these technologies have been used to overcome many of the inherent limitations of the current designs.

As users generate greater load and use more time-sensitive applications, services costs are driven higher. This has been solved by the operators consolidating network infrastructures in their attempt to control costs, but the scaling of load onto a few or a single infrastructure(s) has created service visibility problems for network management. With this infrastructure and service consolidation, the flows disappear into larger and more aggregated pipes, and are therefore are more easily impacted and, when resolution is needed, are more difficult to identify.

For full service management, the following components and concepts have been identified as being required for the management platform of the network. These include alarming, performance monitoring, threshold monitoring, load monitoring, fault diagnostics, cognitive response engine, LNMF, link resource optimisation, element resource optimisation, SLA monitoring, service monitoring, capacity trend planning, and others.

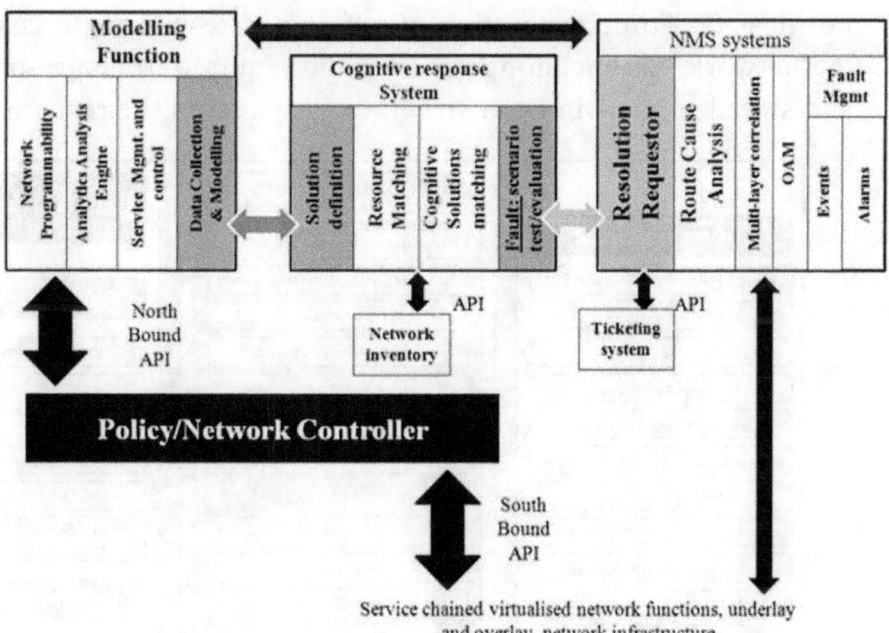

Figure 45: Command and Control functional components

These functions are required to ensure the network can be tuned to automatically adjust connectivity; therefore ensuring the service SLA or service requirement is met. For this to occur, the network and service information must be made available at the different

185

points of control to ensure that checks can be made against the end-to-end service delivery.

Analytics gathered at each layer should therefore be collated for historical and current analysis, and this data should be structured using a Big Data function. Having access to this information will both aid the evolution and support the drive to achieving the automation of service and infrastructure management. This data will enhance the level of intelligent decision-making, required to enable accurate fault resolution.

For this to occur, the monitoring, management and analytics gathering from the infrastructure needs to be structured; therefore allowing for a simpler form of processing. This demands that the command and control systems have access to the structured Big Data analytics gathered from the network infrastructure and the services.

For near real-time generation of automated responses to issues on the network, the incoming data will be required to be pre-processed and stored in the Big Data structure to speed up the response time.

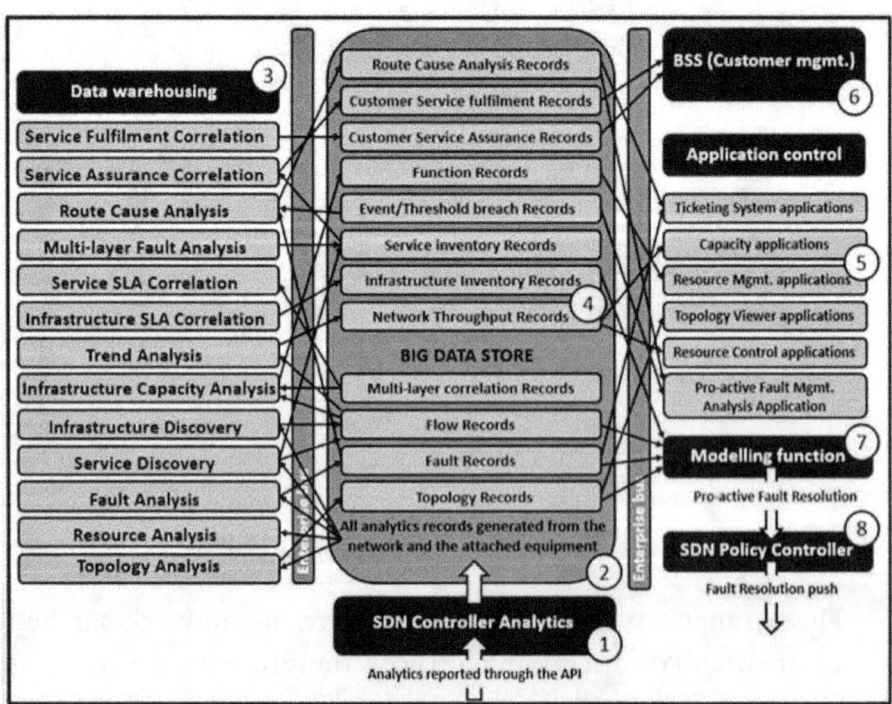

Figure 46: Application analytics data structuring

This also avoids valuable data being stored in proprietary management and business applications. Through the structuring of data for management systems - rather than the collection of data in standalone systems - the business retains the original data. This approach permits reuse of data by other applications (or business units) at a later stage. The value of the data is increasingly being realised by businesses, as is the demand for accurate data across all departments.

SDN delivers a centralised point of control, but still gives permits for distributed and localised management in environments such as vCPE environments, SDN VPN etc. The localised control has the ability to affect the local relevant decisions. When utilising resources northbound of the immediate environment, the local solution uses the rules sets of the centralised control. This means that the central controller doesn't have to address every command into the network environment.

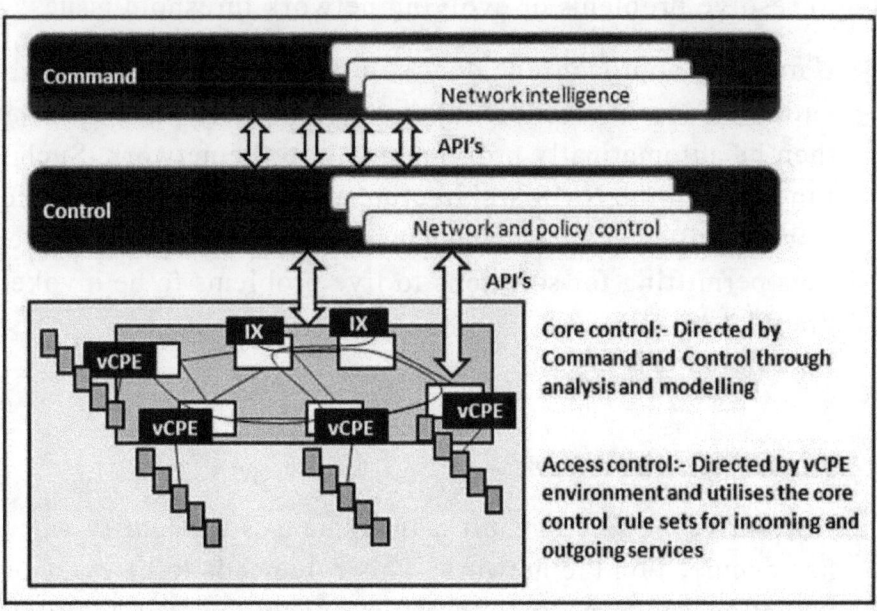

Figure 47: Interworking core and access control

The core network rules sets are controlled by the centralised command and control. This mechanism provides a centralised point from which analysed and modelled data is sufficiently informed to be able to manage the end-to-end core network environment.

187

This allows automated tools to manage and influence the distributed nodes, while providing them the Segment Routed flows for the delivery of an aggregated bandwidth of traffic for differing services types that require extra management and adherence to stricter SLAs. Both traditional network operators and OTT companies can make use of this mechanism to co-operate and to leverage current Net Neutrality legislation – for the benefit of the customers, who are then able to select Quality of Experience for a group of IP flows that make up a single service.

When considering this in the context of the SDN network controller's ability to program the network, significant new capabilities arise for both Internet operators and users. The introduction of the LNMF and its ability to intelligently model the impact of an infrastructure or service fault provides the move to proactive resolution of networking issues. In conjunction with a CRE function, end-to-end polices can now be identified and applied to resolve problems or evolving network threshold issues.

Using this approach, incident solutions can be automatically validated across multiple layers of technology, and the changes can then be automatically programmed into the network. Such new functions in the NMS architecture would allow for the enforcement of corrective network management capabilities to be implemented, thus permitting for solutions to live problems to be invoked at a granular level.

Fault Management

Currently, the role of fault management is to identify when a fault has occurred on the network. This role needs to be extended to detecting, isolating and identifying anomalies and faults in the network and service. In the SDN Strategic Architecture, this is achieved through the monitoring and correlation of events and other forms of error notification across the multiple layers of the network. Problems can also be identified using the LNMF to model the end-to-end network, and to use the analysis produced by these models to identify anomalies or trending anomalies. In a next step, these are reported to the NOC staff, or identified fault use case policy logic is automatically triggered to correct the evolving situations.

Fault management is delivered through the OAM (Operations, Administration and Maintenance) layers. The Operation function is delivered by the NOC teams who monitor alarms and events that were raised; with the aim of identifying issues that affect the service and infrastructure. The administration (resource management) function keeps track of resources, how they are utilised and where they are assigned in the network. The maintenance and provisioning systems deal with upgrades, patching and the implementation of changes on the network.

The Command and Control solution moves forward the fault management function with the aim of changing network management systems. The goal is to move from having OAM systems which notify the operators that faults have occurred, to having a Command and Control system that is automated, proactive and which initiates fault resolution. To achieve this requires accepting and acting upon error detection notifications, the identification of faults, multi-layer correlation of technologies, identification of services affected, validating solutions, breaking down and diagnosing the faults, resolving faults and reporting error situations.

Some solutions already support varying levels of multi-layer correlation and root cause analysis reporting. Through the SDN Strategic Architecture, open APIs give greater access to new levels of analytics. With greater analytics comes the opportunity to create new systems to process this data and to create more network control solutions.

Moving to a Command and Control-based infrastructure requires the addition of new functions. These include a Root Cause Analysis correlation system that is capable of viewing across multi-layer network infrastructures and services. In addition, this solution requires the development of a Resolution Requestor system. The goal of this system is to correlate outputs gathered from the network by the fault management function. Solution requests can then be transmitted to the other new Command and Control functions, e.g.

Cognitive Response Function or LNMF for analysis and further investigation over an API.

This enables a strategy that creates proactive network management, in which application processes can be monitored, capacities can be trended, quality of the forwarding can be analysed, signal strength measured, deterioration of signal identified, etc. Many of the faults that occur on the network are repetitive and have similar, if not the same, signatures - these commonalities can be identified in use cases.

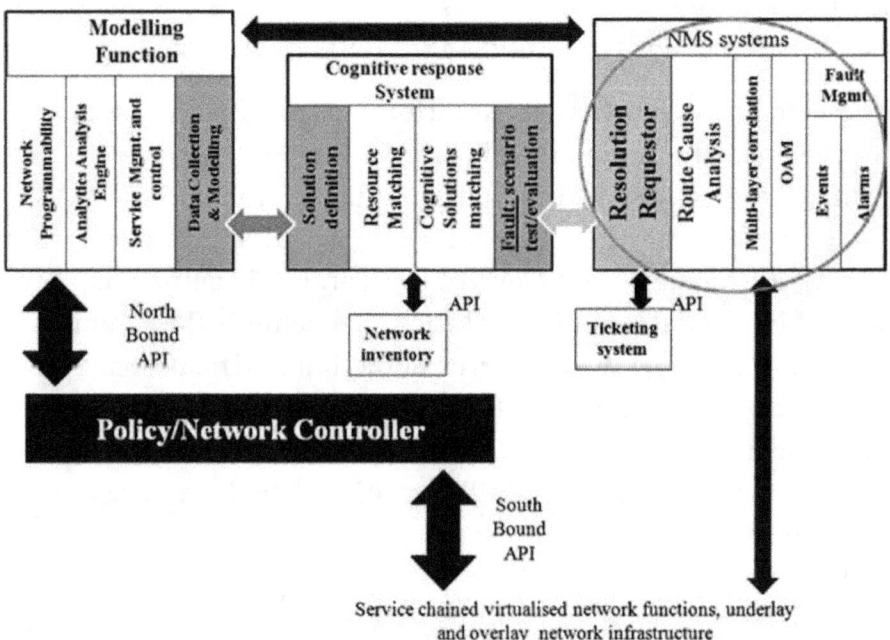

Figure 48: C&C Fault Management System

By using the analytics gathered, the use cases can be defined by operations and engineers to categorise common faults. As time progresses and with additional development, more complex faults can be addressed. This drives a focus on OPEX control. When use cases have been developed for faults and templates created for programmable resolution, the controller can then be initiated by the operations team - or automatically, as the organisation becomes more comfortable with this change in approach. This approach applies the same principle that many operations professionals have always tried to implement through their own thinking when

190

automating tasks through scripting. However, the change that this approach brings is that the new systems can compute and validate the effects of the change, rather than just create modifications using scripts, without the ability to validate all relevant parameters.

The goal and aim of this concept is to move away from fault resolution through manual intervention and to move towards solving complex issues through proactive network monitoring and programming. This approach will support the NOC professionals in their role and supports a transition to a DevOps operations model.

Cognitive Response System

With the introduction of the C&C network management architecture and its integrated network management capabilities, new approaches to automation of fault resolution become available. Driving the automation of fault resolution requires a system that delivers a cognitive response function (This is an author defined term and it refers to a system which takes receives an action, considers the parameters and then creates a reaction).

This function can then investigate faults and provide identification of resolution scenarios that the LNMF can model and validate in near real-time. Within such a solution, a significant number of common fault scenarios can be pre-programmed and pre-validated, which is why this approach also permits for the real-time identification of intelligent solutions.

Such a functional system would significantly change the operating environment of the NOC. If an event was to occur with such a system being available, the NOC team would receive suggested solutions - instead of receiving a multitude of individual multi-layer alarms and events, many of which are misleading in relation to the actual fault. Initially, simple scenarios will be addressed. These could automatically be made available to the NOC teams for final validation. As trust in the solution would grow, for common problems, the appropriate policy-controlled template-based use-case configuration solution could be triggered automatically through the policy controller, and via the SDN controller. This will permit for automatic fault resolution.

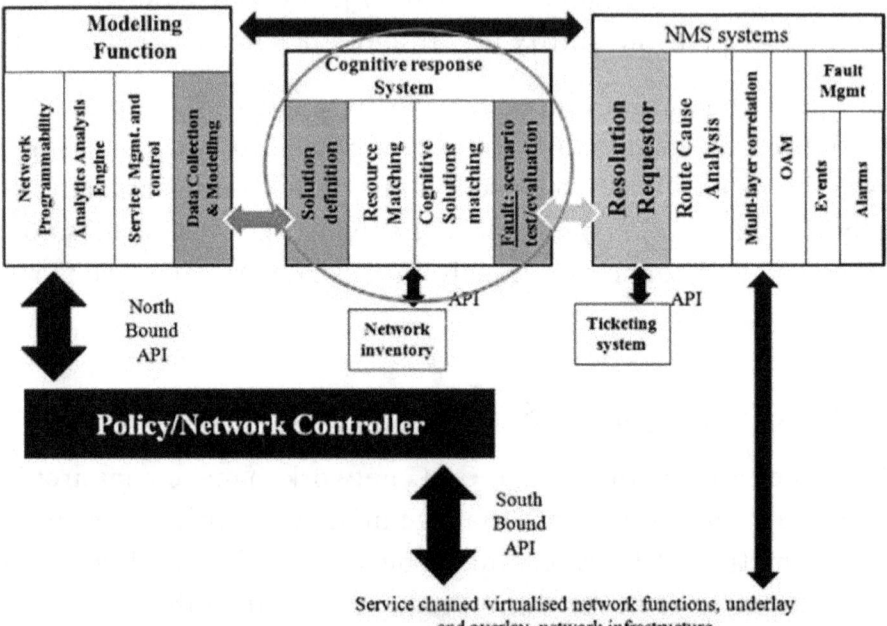

Figure 49: C&C Cognitive Responce System

This Cognitive Response Engine sits between the NMS fault function and the LNMF. It hooks via APIs to the network inventory solution to permit validation of resources. The aim of this solution is to analyse the requests from the fault management system - to support the Cognitive Response Engine in its role of identifying possible solutions.

Some vendors already support varying levels of multi-layer correlation and root cause analysis reporting. When faults are identified on the network, the Resolutions Requestor function can, based on the aggregation of the events gathered from the network from the NMS system, request from the Cognitive Response Function (via an API) a scenario investigation. Possible solutions would be identified in the cognitive solutions matching function, and then resources queried against the Network Inventory system.

Next, the selected solution would be defined, and a full analysis of all impacted endpoints would be correlated. The general programming requirements would be identified, and an appropriate

192

policy determined to affect the fault resolution. Based on resources evaluation, the LNMF would model its final validation of the solution. In a next step, the LNMF would either trigger the change automatically to the network via the API, or would have it validated by the appropriate engineer depending on internal company processes. The SDN controller would then abstract the generalised instruction set and it would be programmed into the specific devices types, within the network to resolve and address issues across the environment.

For SDN and NFV to operate with full efficiency, the network-wide state needs be known, as the controller or modeller can only be as accurate as the real- time information they receive and process.

Furthermore, NFV (Network Function Virtualisation) brings the ability to generate a greater number of functions within the network than Operators have experience of. In conjunction with the ability to spin up new virtualised nodes, this adds stress to the network management environment, which requires greater focus and priority for network management systems in organisations. This, in turn, will require greater scaling of the management solution, awareness of the location of these nodes, orchestration of the virtualised environment, and programmability to meet the needs and speeds of the end user or application.

As with Cloud and SDN, NFV, the network management Command and Control system will have its own orchestration system that it will use i.e. to control the allocation of its resources and to understand the resources on the network.

For addressing the network needs of NFV, SDN introduces a real-time network resource monitoring and management solution. This systems function includes the topological map, real-time recording and accounting of statistics and analytics, and real-time resource management and control. These additional systems hook into the Cognitive Response Engine, providing the full network information to allow it to resolve and analyse problems on the network, and to send appropriate modelling resource information through to the LNMF.

The topological database maintains a record of the active state of all the network resources. This database records the multi-layer network capabilities, which includes visibility of the underlay (physical optical, IP, Ethernet etc. infrastructures), the overlay (logical assigned capabilities) topologies, and the resource available or allocated.

Real-time accounting records in statistics covering all layers of the topology which are relevant for day-to-day management, such as throughput, OAM, and others.

Resource management records and tracks infrastructure and service layer resource utilisation across all layers of the network. These records are used to support other Command and Control queries for the positioning of new services, and to support the generation of Modelling function queries.

Tenant Management controls the multi-tenant authentication and access rights management to network resources. This function is required in the Network Management system to ensure that resources are identified and can be delivered when a service requires them.

Network Services Inventory maintains the real-time state of the networking infrastructure and services to ensure that an accurate reference system exists. The strategy of having the systems is vital to ensuring services are delivered at a granular level, and to make sure that any constraints on resources are known, should a new service be requested. Having this information available enhances the capability of both proactive management of services and infrastructure, and Root Cause Analysis generation. These are all building blocks that enable and make sure that the customer receives the service according to with their expectations. It is these solutions with their enhanced monitoring capabilities that can enable this change in the approach to network monitoring.

Live Network Modelling Function

A key principle of SDN is to generate higher utilisation of the network and to bring about greater utilisation and cost-efficiency in networking.

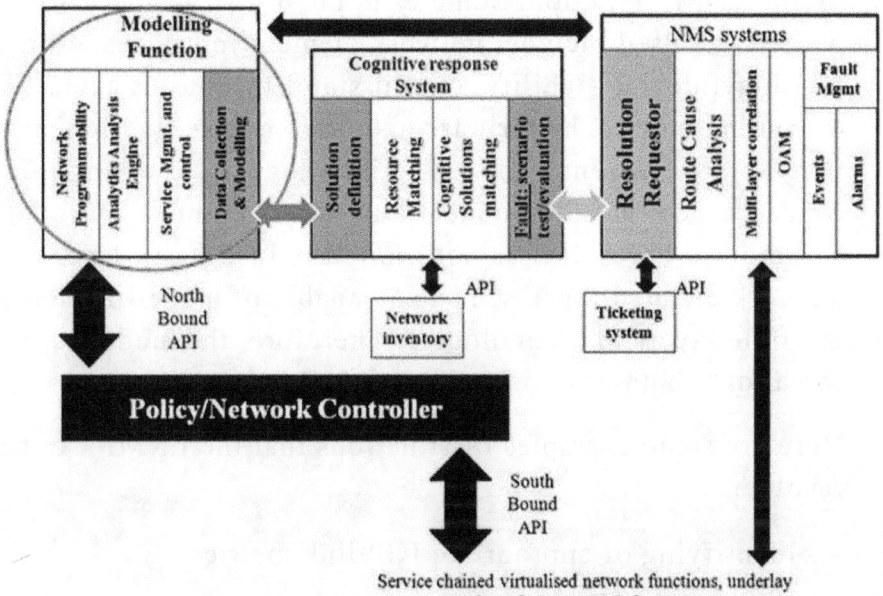

Figure 50: C&C Live Network Modelling Function (LNMF)

For optimising the end-to-end infrastructure, the LNMF, modelling capability can be used to make sure that the network is optimally tuned. In today's bulk forwarding IP networks, inefficiencies exist due to the router being the sole decision-maker as to how traffic will be forwarded. This occurs because configuration parameters are set without an end-to-end view of the services, leading to inefficient programming and utilisation of the network infrastructure as the resource demands change.

The LNMF is not new in IP Networking; these solutions have existed and have been used in operator networks for more than a decade. Historically, they have been offline systems that have been used by many operator organisations of various sizes to support the design work, capacity planning, costing, budget support etc. Additionally, they have been used to enable a higher level of operational stability on the networks by advising on better metric parameters to use to program the network.

195

Scenario Modelling

The LNMF systems are able to model extremely complex scenarios at a flow-based level, using statistical flow inputs from Netflow, IPFIX, JFlow, SFlow. Many other inputs are also collected and used by the LNMF function. Some example of the hundreds of parameters used include: network element interfaces, circuit topology, device stability, circuit stability, routing tables, stability of routing entries, historic trends, circuit capacity, optical inventory, IP inventory, peering interconnects, peering interconnect capacity, peering interconnect load, peering interconnect stability, and others. The LNMF is not limited to just the IP layer or the IP network element layer; it is also capable of modelling across multiple layers of technologies. Therefore, the models can be used to validate both designs and operational situations.

Here are some examples of situations that the LNMF can be used to resolve:

- Identifying of appropriate IGP link metrics

- Validating the RSVP TE tunnel set up is correctly implemented

- Identifying risk groups on fibres

- Validating flow-based load issues

- Validating trend issues on networks

- Identifying where congestion can occur as the growth trends occur

- Identifying current and future SPOFs (Single Points Of Failure)

- Identifying load issues on routers during failure scenarios

These are but a few of the many hundreds of scenarios that these systems have been used to resolve in real-world networks for over a decade.

Offline versus SDN LNMF

Such systems have historically been sold as standalone solutions. However, as these systems have been offline, they do need an architectural change to be able to operate in real-time/near real-time. With SDN Strategic Architecture incorporating the modelling function into the operational running of the network, this function now no longer remains offline, but becomes a real-time programmable network management element. This capability enters the architecture at a critical moment in time. With the increasing network complexity causing continual OPEX growth, and the desire of many companies to move from the traditional operations model to a DevOps model, this tool provides a valuable addition to the operations team when faster fault resolution is required.

In the SDN Strategic Architecture, the LNMF system is considered as a multi-layer modelling, management and control solution. With the network modelling functions already having the ability to do multi-layer modelling across the fibre path, the wavelength, the optical node and its interfaces, the router and its interfaces, the flow of traffic from its point of origin to the BGP next hop, and peering interconnects, the incorporation of this functionality brings considerable management capabilities to the network. These capabilities are significant; they give the ability to map services with their SLAs to the network infrastructure using Segment Routed paths. This allows for the separation of services across network infrastructures according to aggregated services needs.

Optical

For the optical layer the multi-layer modelling, management and control solution brings new functionality possibilities. When considering the optical evolution, the opportunity arises to enhance the capabilities of GMPLS (or other such protocols) and to provide a dynamic optical capability to the operator. When a failure occurs, the modelling function can be used to identify an alternative optical path, therefore ensuring that the customers' service is maintained even during an outage. When an alarm occurs, the LNMF can simulate which alternatives exist and then, using protocols such as GMPLS, identify an alternative path and have it triggered into the network.

As most networks are relatively fixed in structure, the LNMF can pre-compute alternatives, therefore speeding up the process or solutions identification. The goal of this work is to achieve the end-to-end service capability that maintains the customer SLA. This drives down costs, as capacity no longer needs to be locked away on optical networks and enables consideration to be given to new structural architectures. This drives a new cost model for the network and improves network management, as some optical outages will no longer constitute a significant impact to the network.

For the delivery of an enhanced and complete optical reconvergence decision-making capability, it is necessary to address the exposure of the optical topology and interface information from the proprietary optical solutions – otherwise this will drive a design that creates a Modeller-Of-Modeller controller function that will constitute a loss of visibility and lack of end-to-end service control. It is, however, likely that the first step in enabling dynamic optical will be the Modeller-Of-Modeller capability.

Optical vendors are co-operating both with each other and with operators to identify a standard that will expose the topology and interface capabilities within the individual vendor network topologies. This will create significant end-to-end service visibility for network operators by reducing the complexity of running multi-vendor optical infrastructures.

Modelling "Intelligence"

Use of the LNMF changes the approach to control and management of a network and supports the Operational teams in gaining operational visibility of the network. Through the use of the LNMFs enhanced network topology computation capabilities network infrastructure management can be enhanced in ability and reduced in complexity. These capabilities further enable the move from infrastructure management to service management. This creates the opportunity to compute against failures, new service impacts, change on peering impacts or trends on a network-wide multi-layer scope.

Though the additional gathering of analytics on the network infrastructure and the possibility to do offline path computation and resource optimisation, the SDN Strategic Architecture exposes the necessary visibility and topology information to perform service management on a per-flow basis.

During incidents on the network, this important capability enables the NOC team to automatically review and re-optimise any impacted services. Service flow modification can then be reprogrammed into the network, based on predefined policies, using the generalised northbound API to the network controller.

With the capability of node and service monitoring, comes the need for considerably greater granularity in statistics being gathered from the various nodes and service points across the network. LNMF systems are able to gather the required analytics information, but this data should also be replicated to the Big Data store to ensure that the data is available should any offline evaluations be required. LNMF creates the opportunity to manage not just the infrastructure, but to also manage and program the network to react at an application level.

With the LNMF permitting for a new level of visibility in the network, the NOC environment is provided with a near real-time understanding of the network resources.

Layer 2.3 - SDN Real Time Management and Control

An overview of some of the components in Layer 2.3

- Command and Control – Real-time Capabilities

- Policy Controllers (Service Provisioning)

- SDN Controllers

- Southbound API (Protocols)

Layer 2.3: Functionality Description

This sub-layer of Layer 2 operates in real-time and triggers configuration changes based upon instruction sets received from all layers. The focus of this layer is to control the network via Policy/SDN controllers. This layer uses the LNMF to provide proactive network management capabilities. Configuration changes are driven into the network through the southbound APIs.

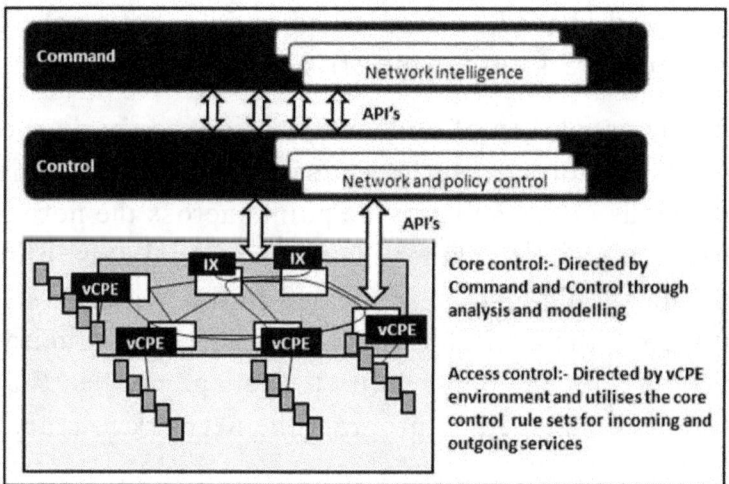

Figure 51: C&C - Policy enabled intelligent fault resolution

Sub-layer 2.3 operates in real-time and triggers configuration changes into the network through the southbound APIs based on instruction sets received from all layers. The focus of this layer is the network and service control using the policy and network SDN controllers.

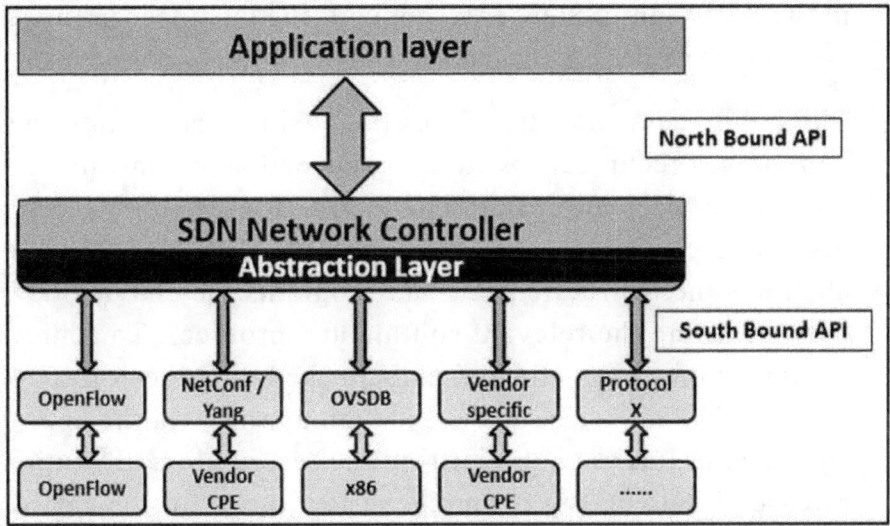

Figure 52: Application based Real-Time network and service control

These functions focus on the control of the network and the configuration of instructions into the network. This way, NFV and Cloud orchestration systems can initiate and request the creation of network resources to complete the automatic delivery of the end-to-end service solution. These systems allow for generalised policy-based instructions to be received through a northbound API from a variety of systems. These generalised instruction sets are then applied through the SDN abstraction layer to the end device, irrelevant of the vendor CLI.

Policy-based Control

SDN with its Command and Control approach targets the strategic development of an architectural solution that will give a full view of the network state, and that permits for the use of policy controller to both reconfigure the network state for new services and to address problems within the network environment.

This architectural approach permits policy control to be effective within the SDN environment. There are many network components and environments within the network where policy-based configuration can be used for reconfiguration, such as the network elements, the vCPE environment, control of fixed and mobile access

201

networks within a SON environment, the customer premise equipment or the end device and applications.

The policy controller translates the business needs into the generalised technical instruction sets and communicates this via an API to the SDN controller. The instructions identify the service needs and parameters, and from this, the SDN Network controller abstracts the instruction set, and programs the end device in the network using the relevant southbound protocol. To achieve this, resources are first validated and checked, and the correct device is identified. The instruction set is then applied via the appropriate southbound interface, permitting the reservation, allocation or release of networking resources. This allows for the networking resources to be allocated correctly to the higher-level service.

Attaining a simplified approach requires a per-function template approach to configuration in order to support the creation of a customisable service and its application onto an appropriate device. Since the user no longer wishes to be constrained to a fixed point with their technology, a shift in the network configuration model is essential to support the evolving needs of the consumer.

Simplified provisioning is one of the main aims of SDN. It targets at automated and simplified service creation across both the underlay and overlay. Policy-based control permits for a workflow approach based on service templates that allow for the instantiation of accurate customer and infrastructure configurations, irrelevant of end vendor configuration CLI language. The historic problems of achieving a simplified approach to provisioning highlight the complexities and costs incurred by the network operator when required to use proprietary vendor CLIs. It also shows the benefits of open approaches, such as OVSDB or OF-Config, which can drive significant operational and design simplification for the operator. Such solutions, may not yet provide all the features that are required, especially when compared to what many in the industry are used to having available in vendor proprietary solutions, however with further support from across the industry this does not have to be the case.

When using the SDN Strategic Architecture, the policy control function sits within a hierarchically layered structure. This creates the ability to interface to other functional systems, within the scope of set rules, and via an API. This way, programming instruction can be fed in from the service catalogue, enabling order provisioning or order changes. It permits for proactive network management capabilities to be structured and programmed into the network using rule based templates.

This level of structure and reusability creates an architectural approach which is comparable to object-orientated in structure and which permits classes of objects to apply methods using common interfaces. This approach drives a policy-based abstraction function that permits for the creation of controls on a product, application, user, group or security control levels; all through a common interface. These templates can then be communicated through open APIs, and therefore permit for an open and generalised approach for the application of configurations onto the end device.

Policy controllers can apply a fixed logic based on templates. The term 'policy' has a very broad meaning and could be defined as; a set of generalised predefined rules which when applied create the intended state required for the purpose of delivering a service. The advantage of using a policy defined template to ensure the intended state is applied is that when the intended state does not match expectations, non-compliance alerts can be generated. This can be used to generate the appropriate alert to relevant systems and teams, and to identify the problems that exist with the service. This fixed template structure also works well for analysis and diagnostics through the LMNF and CRE, as expectations on what the correct configuration template should be are pre-set.

Policy on the network encompasses many configurations and touch-points, and templates are used to centralise policies for ease of use and reuse. Policy controllers, in conjunction with network controllers, are used to automate provisioning of service and infrastructure, therefore ensuring error-free provisioning and consistency of deployments and services.

Template-based policies avoid the need for repetitive generation of command line for each of these functions by using network policy templates for functions such as access lists, interface configurations, ISP rules definition, service rules for vCPE environments etc. This allows for the 'design once, and apply many times' principles. These principles are the same as those used in OOP, where reuse rather than re-write of code ensures that less testing and code validation is required. As with OOP, the principles of single purpose methods is expected when creating policies, therefore ensuring that there is clarity in the application of the template building blocks.

Northbound API

A full chapter is dedicated to the northbound API in this book (14). To avoid duplication, only a description of where the function of the northbound API sits in the SDN Strategic Architecture is given in this section.

Key to the SDN Strategic Architecture is the ability for applications to communicate in a standard language and with a standard instruction set, irrelevant of the equipment in the network. With SDN having introduced this IT systems concept into networking, the SDN Strategic Architecture utilises this structure to create layers within the architecture. This grants flexibility in business decisions regarding the selection of vendor equipment in the network.

This functionality is provided through APIs. These APIs enable operators to continue leveraging the investments already made in the current infrastructure. They also enable applications to access networking resources for the purposes of consuming, manipulating or querying the state of such resources, and they are standard functionality used in the IT side of the business for many years.

One of the most important northbound APIs is the REST (Representational State Transfer) API. It is not a standard, a protocol or a programming language. The REST architectural style was developed by W3C TAG (Technical Architecture Group) in parallel with HTTP 1.1. It is based on the existing design of HTTP 1.0. REST is a natural fit with the SDN controller solution as it provides the ability of the communication of specific generalised instruction sets. Other APIs are being considered for use within the SDN architecture. As REST uses HTTP, there is a wide set of experienced people with the right skills to develop and to create the APIs within and around the industry.

The northbound controller REST API interface is used for the communication between the controller and the applications or business logic. This works well with an SDN Strategic Architecture, as this forces layering into the structure, thus permitting for a single generalised approach to communicating the instruction set to the controller.

REST focuses on the roles of the components, the relevant data structures and the REST constraints that need to be considered when interacting with other components. By focusing on these and not on the details of the component implementation or its component programming, REST allows for a single generalised instruction set to be communicated. This can later be abstracted to the appropriate component level and protocol syntax by the relevant controller systems. The constraints that have been applied to the REST API architecture ensure that any development is done are in a scalable and controlled fashion.

SDN Controllers

A chapter is dedicated to the SDN Controller in this book (SDN Controllers). To avoid duplication, only a brief description is given here of where the function of the SDN Controller is and what it does.

The SDN Controller is at the heart of the architecture. The SDN central controller solutions allow network operators to deploy an intelligent, software-defined control solution. This SDN controller provides the capability to interface between a C&C system for active management and the infrastructure for services management across multi-layered networks.

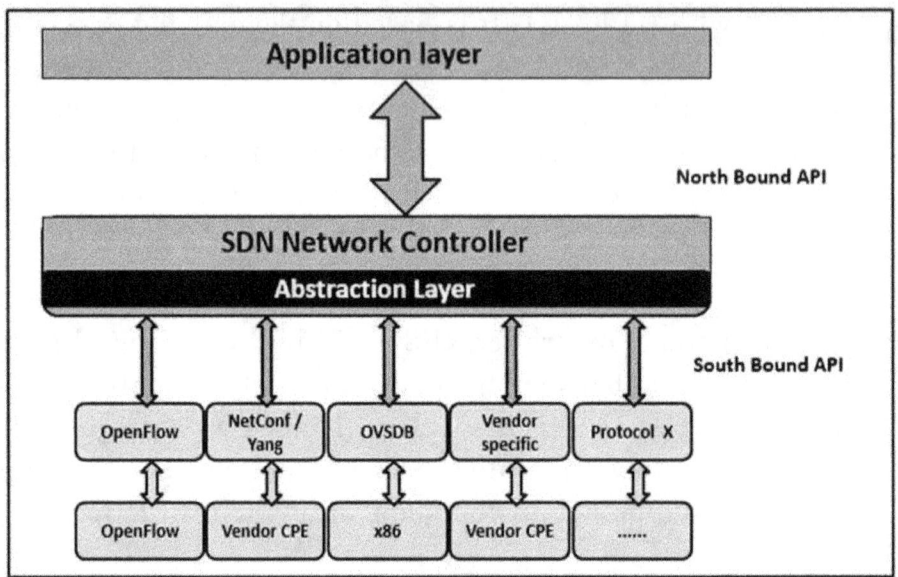

Figure 53: SDN Controller functions

In most evolving SDN controller solutions, this functionality is very intertwined and not yet separated into the necessary layers of near real-time and real-time systems. This separation is required to ensure the authority of the SDN Strategic Architecture - it avoids an architecture evolving that would permit multiple systems having the authority to make changes across the infrastructure. This structure makes sure that the setting of control instructions into the control plane is limited, therefore conflicting control messages being injected into the network can be avoided. Architecturally, these systems are delivered with redundant solutions.

For clarity, the OpenFlow™ protocol is not all there is to SDN. OpenFlow™ is one possible option for enabling SDN solutions, but other SDN controllers and protocols are evolving as more networking requirements are considered. These SDN controllers provide the abstraction of generalised commands to other protocols as well as OpenFlow™. These additional southbound protocols/APIs are used to communicate relevant instructions set to the end device. The programming of the end devices via the southbound protocols are addressed in layer 3.

11.3 Layer 3: Service Aware Network Infrastructure

An overview of some of the components in layer 3

- Routers control, configuration and management

- Optical equipment control, configuration and management

- Switches control, configuration and management

- Protocols: Segment Routing, IP Meta Data and Network Service Header – Control and Configuration

Layer 3 Functional Description

This layer utilises the southbound APIs to initiate product changes based on pre-defined programming instructions, templates or data models, depending on the architectural protocols defined. Analytics gathered and received through the southbound APIs are utilised locally by the SDN Controller to generate a response, or are fed to the appropriate higher-level system. The higher-level system will record, analyse and, where required, the LNMF, the NOC or the policy controller will generate a response to trigger a change into the network.

Southbound Programmability

Chapter 15 is dedicated to southbound API/protocols in this book (Southbound Protocols or APIs). To avoid duplication, only a description of how the function of the southbound API/protocols exists in the SDN Strategic Architecture is given in this section. Network southbound APIs are a set of southbound protocols that are used to hook into end switches, routers, etc., to set configuration parameters to control and manage the device either by programming the forwarding table or by defining the configuration.

At this point in time in the evolution of SDN, many new protocols are being created, as vendors open their existing architectural solutions to permit for an SDN Controller-based architecture. These southbound protocol/APIs enable the control of the network resources through a centralised controller.

The northbound API receives generalised instruction sets. Once the SDN controller has validated the constraints of the end device, these are abstracted by the SDN controller via the relevant southbound protocol/API. Next, the instructions sets are issued through the southbound API to configure the end device. The commands issued through the southbound API define the configuration for the end device and forwarding, security, and other rules. Within the context of the SDN Strategic Architecture, the southbound protocol/API is required to be capable of returning network and service statistics for analysis and service reporting at a higher level. The statistics are analyses to validate the success of the device configuration and the state of the services running through it. Analysis is be completed by applications residing in the real-time and near real-time layers north of the SDN Controller.

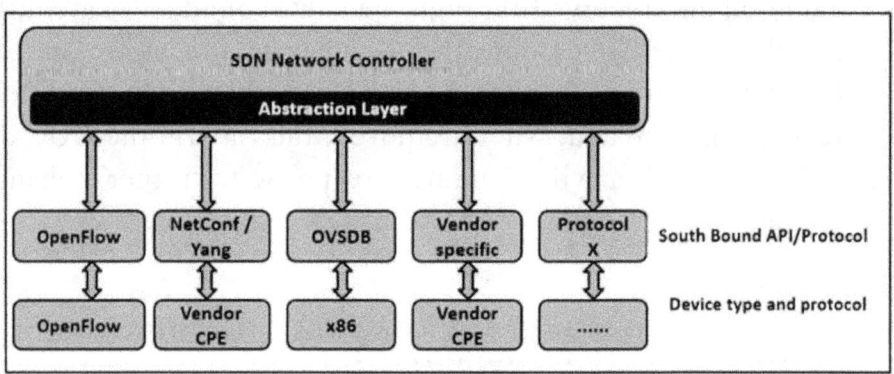

Figure 54: Example of Southbound API/protocols

Network Programmability via southbound APIs (protocols) allows for the ability to expose the control of the service. This is one of the fundamentals of automating the network control; it is also one of the fundamental breakthroughs that the SDN brings to the future management and programming of multi-vendor environments where proprietary CLI still exist. The programming permits for control instruction sets to be activated on the control plane of the underlay network.

At the moment, with so many vendors moving forward to SDN and opening up their once proprietary systems, a lot of southbound

APIs/protocols have quickly evolved. This number is expected to reduce over the coming years as vendors apply the full extent of the programmability, get a better understanding of the operators' needs, and gain the benefits of standardisation. Some southbound APIs are protocol-based and require end device support of the protocol on the device to permit for the device to be programmed. Others use a table-based methodology and translate instruction sets to the historic CLI in use by vendors. These different approaches resolve different issues for operators and permit the current vendors to make a rapid move to SDN. Each approach has its benefits, but it is expected, as these solutions become more commonly used, that the feeding through of operator requirements will identify solution smaller number of south bound protocols.

Depending on the vendor approach engineers may define the configuration using a configuration snippet, data model or GUI template based approach. These solutions can be considered to be methods within an object-orientated approach to delivering of solutions. These snippets are defined one time for each service and infrastructure requirement, and are then called and combined to create the total configuration solution. Resources are managed in the resource systems, and these allocate unique resource identifiers to the individual solution.

11.4 Summary

The SDN Strategic Architecture is as much about enabling structured working practises as it is about sharing information on how SDN can be used to create a business-focused architecture.

Key to any architecture is the working practise that comes with the architectural approach. This architectural approach enables the multiple teams, all working on individual areas of the end-to-end architecture, to control their own domain, while having a clear understanding of what is required when interfacing with other teams.

Using APIs, SDN technologies structure the architecture into layers, and the SDN Strategic Architecture enhances this into the higher layers. In accordance with their original purpose within the IT systems architecture, APIs create a definitive demarcation line between systems and provide an interface point between the different groups and their area of responsibility. The demarcation line helps structure responsibilities and enables interworking between the different teams. It also supports interworking between colleagues within the same team who are required to deliver different functional components.

The API interface point makes sure that different groups can clearly define and specify their expectations on what is required to ensure project delivery. This is achieved because layering provides clarity on the technical complexity of the deliverable, thus enabling the technical professionals to highlight the detailed information that they require early in the process. This supports the technical project management team who can make the essential information available, thus helping to identify the state of the project deliverable. The SDN Strategic Architecture has been structured to strengthen successful practises used across the industry.

With an API store model already becoming available, architects can access commercial "API stores" and identify, purchase, or lease APIs for use within the architecture. This reduces the development time and drives faster time to market for network operators.

The API store model can be a website on which individuals and companies can register APIs they have created for sale or for lease, enabling others to purchase or lease an API and to avoid API development.

The availability of the API store model aims to speed up project delivery for the network operator, while enabling individuals or companies to utilise investment they have made to generate an alternative income.

12.   Object Oriented and Agile

The SDN Strategic Architecture loosely borrows some of the structuring used in an object-oriented software development methodology.

The development of the SDN Strategic Architecture has been loosely based on this methodology of structuring systems that have their own function, process their own data, assign responsibility for individual behaviours and do this through using an object- and method-based approach. This concept has not yet been fully investigated or fully developed, but it is discussed hoping that, through sharing these thoughts, others will consider expanding on such an investigation, should they consider the ideas to have merit.

Such a finished and defined methodology would be invaluable to the network operator and to vendors. Staff could be trained for a better understanding of the solutions' complexity, and each function would have clear architectural roles and responsibilities.

Anyone with experience in product development has seen the expansion of functionality within a component. Some of this has been requested by the customer, but much will have been done for competition reasons by the vendor. In some cases, the total sum of additional features added to a platform can change the function of the platform and can lead to the creation of an architecturally unsound system. This can lead to significant scaling constraints, depending on which features are actively enabled at a point in time.

By using an object-oriented functional block approach, this would enable architectural control to be placed on functionality or feature creep. If a methodology was used, the trained staff would be able to quickly validate if the proposed approach should be delivered in a separate functional application or as part of the current application. Such a structured process would significantly aid development teams, product creation, vendors and the operational teams.

Object Orientated Term	SDN equivalent	Object Orientated Description
Class	Service catalogue	Classes are composed from structural and behavioural constituents. They are concrete data structures.
Class	Network element	
Class	LNMF	
Class	Customer	
Object	Customer preferences	Objects are created as instances of a class. This provides initial values for state and implementations of behaviour.
Object	OVS	
Object	Firewall	
Object	Router	
Type	Open Flow	Type is an interface that allows unrelated objects to communicate with each other.
Type	XMPP	
Type	NetConf-Yang	
Method	Policy Controller	The behaviour of a class or its instances is defined using methods. Methods have the ability to operate on objects or classes.
Method	Network Controller	
Method	Product policy	
Structure	Security rules	A class contains data structures that can be associated with variables, which either belong to the class or specific instances of the class.
Structure	Forwarding policy	
Structure	Configuration snippet	

Figure 55: Mapping object-orientated functions to SDN functions

This would be especially valuable for the operations teams when analysing complex faults, as an identifiable structure could then help isolate the source of a problem. Such an approach is very appropriate in a NFV environment where scaling of virtualised functions can be problematic. Singular function based systems will be easier to architect, scale and manage in the infrastructure than multi-functional components using developing (NSH) service chaining standards.

This brings additional benefits as it helps identify the responsibility of the function, permit encapsulation of the data, allow for inheritance and ensure for reuse of development work already completed.

If during development an aspect of the solution failed to meet expectations defined by an object-oriented approach, this would highlight questions about the architectural development. Having expectations on the systems, applications and component functionality could help create guidelines for product development and products sourced from other teams or parties. This would enable the teams delivering the development to clarify which component or enhancement was needed in which system.

This is not to say that external features cannot be absorbed and utilised for the sake of speed. This suggested approach is about understanding and ensuring that, when change is made to a platform, the implications to the end-to-end solutions architecture are identified and understood.

It may be necessary to incorporate the functionality in the short term, but for the long-term solidity of the architecture this may need to be moved to a new platform. This would provide guidance on boundaries to those delivering the applications or components, and it would encourage them to develop other applications as part of an end-to-end architectural approach.

In turn, this could limit duplication of functionalities across vendor-sourced platforms and provide rules to the network operator as to when platforms are no longer of value in the architecture.

12.1 SDN: An object-oriented approach

Architectures, when considered within the historic component-focused structures, were based on proprietary systems and reflect the limitations that these proprietary systems created for the architect or designer. Although many network protocols could be used in the creation of a solution, the lack of data exposure and end-to-end control constrained the architecture from addressing the requirements of the business and its business model. These technology limitations influenced the products that could be sold and the SLAs that could be guaranteed.

By using an object-oriented development approach, the SDN Strategic Architecture can evolve a company-specific system and networks architecture by helping to identify and guide the creation of the systems and functions, tailoring the architecture to meet the companies' business aims and goals.

A structured architectural approach would further help to remove pockets of confusion for other team members, thus supporting those involved to understand how the project delivery executes the technology strategy set out for the company.

Object-oriented design definition

A design method, in which a system is modelled as a collection of cooperating objects, and where individual objects are treated as instances of a class within a class hierarchy.

Four stages can be identified:

1. *Identification of the classes and objects*

2. *Identification of their semantics*

3. *Identification of their relationships*

4. *Specification of class and object interfaces and implementation*

Source: http://dictionary.reference.com/browse/object-oriented+design

In SDN, the separation of the control plane from the forwarding plane breaks the combined functions within routers that up until now have been integrated. This permits for the separation of unaligned functions and the creation of functional blocks (classes).

Such a process of separation can be continued within the OSS stack functions and network management solutions. Through this approach, functional blocks become apparent all across the service delivery process. These include policy controllers, LNMF, CRE, network controllers, etc.

With open interactivity now feasible between systems (classes and objects) through the use of APIs (interfaces), the separation of functionality allows for the necessary application (methods) to deliver the infrastructure and service management for the individual customer products and preferences (objects). This allows for inheritance and reuse of data and applications in the context of service development.

This approach is object-oriented as it permits the network architect to structure the functions, systems, policies, configurations, configuration snippets (templates), etc., into the appropriate objects, classes, types, instances, etc. With the additional flexibility introduced by SDN, NFV, Big Data, Cloud and a NGOSS, the architecture can be defined within the concept of structured building blocks, where key programmability elements can query or call other instances, methods, types and structures.

Object Orientated Term	SDN equivalent	Object Orientated Description
Class	Service catalogue	Classes are composed from structural and behavioural constituents. They are concrete data structures.
Class	Network element	
Class	LNMF	
Class	Customer	
Object	Customer preferences	Objects are created as instances of a class. This provides initial values for state and implementations of behaviour.
Object	OVS	
Object	Firewall	
Object	Router	
Type	Open Flow	Type is an interface that allows unrelated objects to communicate with each other.
Type	XMPP	
Type	NetConf-Yang	
Method	Policy Controller	The behaviour of a class or its instances is defined using methods. Methods have the ability to operate on objects or classes.
Method	Network Controller	
Method	Product policy	
Structure	Security rules	A class contains data structures that can be associated with variables, which either belong to the class or specific instances of the class.
Structure	Forwarding policy	
Structure	Configuration snippet	

Figure 56: Structuring functional component architectures

This thinking is not complete. However, these initial thoughts are shared to with the intention to encourage others to take this thinking further. It is hoped that this will also give architects and designers another tool to use, when considering how to achieve a strictly defined architecture that can easily absorb flexible growth and modification.

Once a function-based methodology that sets rules on the types of interaction permitted between systems has been defined, engineers, architects, designers, and operations professionals can adopt it and learn how such a methodology can be used cohesively throughout the organisation.

This provides extra checksums in the architecture to make sure that future modifications are not in breach of the aims and goals of the overall architecture, and, with end-to-end understanding, ensures that hacks will be addressed. This approach could guide not just the internal network operator teams but also the vendors, when they create functional (class) systems.

Agility of service management and control is achieved through the ability of the controllers to abstract management and control of forwarding, thus permitting the architect of the service to dynamically adjust network-wide traffic flows, to meet the requirements of the customer and the service.

By using an object-oriented approach, this extends the architectural control beyond the functional blocks to the data structure and data types. This way, the data structure becomes an object that addresses both data and functions. Using this methodology, relationships can be created between one object and another, therefore allowing an object to inherit characteristics from another object, and therefore allowing the functional component to become greater than the sum of its parts.

By leveraging the service catalogue function, SDN controllers allow additional objects to come together and to be applied through a common interface. This, along with the ability to create interfaces, use types, and, to invoke structures, creates the ability to define an end-to-end services architecture using a methodology that was not possible with component based architectures.

The object-oriented programming technique of permitting the creation of reusable objects is replicated through the SDN architecture, in which instruction sets can be reutilised in the creation of new services. This ensures greater testing of the components and better reliability in the key designed components.

With SDN and the SDN Strategic Architecture there is an opportunity to address the network and service management, and to structure the architecture into an easily interfaced, open, layered model. This creates the opportunity to have more flexible functions that are open and permit the ability to focus on data and service - thus allowing for flexible, customisable products.

When evolving the SDN/NFV architectural capabilities, any new function could be considered in the light of OO terms. The aim is to avoid a crossover of functionality between different functions and to limit the development of over-scoped solutions. These rules could be used to evaluate new vendor products, to ensure they meet the needs of the individual companies' business goals.

12.2 SDN influence on Agile Scrum and DevOps

Agile Scrum is widely used in the development of complex project development where highly skilled individuals are required to collaborate and communicate.

SDN, by its nature and evolution, further enables product development delivery and design in an Agile structure. It is suited to this methodology because of its open APIs, its functional components and its software focus. SDN and NFV with their functional structure complement working with this methodology. However they further complement it because of the approach as to how architectures and design are created. Unlike the technology silos' waterfall development approach, where one silo of the technology team would 'waterfall' their part of the solution over to the next component team, Agile supports the delivery of key functionalities by using 'sprints', This validates the chosen approach and can then be expanded over time, which allows greater flexibility for products to be scoped and tested in the market without having to wait for full end-to-end development to be completed.

SDN enables functional solutions to interface with each other through a layered API approach, providing communication, abstraction and specification. This use of APIs opens up a development and testing approach that is similar to the principles of 'Definition of Done' and aids the identification of technical and functional spikes. These layers help to define the hand-off between the various functional solutions within the SDN Strategic Architecture.

In an Agile environment, with each piece of development being delivered within an Agile sprint, there is a need to control the interface to the other functions and to ensure that the interface supports the needed capabilities. Agile planning is aided because the Scrum team members responsible for the self-programmable APIs between the functional systems will be aware of the requirement changes needed to modify the product. As these APIs are programmable, the changes required, even between disconnected groups, can be scheduled to be developed in an experience-based timeline.

Within the sprint model, the work should be able to be scoped and identified in advance, and the appropriate teams in charge of the APIs and/or functional control systems pulled in at the right time. This requires that those developing the product to have a reasonable understanding of the key interfacing technologies that are used. The SDN/NFV layering of APIs adds extra structure and, through this, supports the technical development teams by ensuring that proper thought is given to security and other critical topics when information is being handed of through an API.

Through the use of policy and network controllers fed by APIs, an Agile team can avoid rebuilding component-based interfaces on a multitude of protocols. The standardised agile approach to development thus also leverages work already completed: classes, objects and interfaces that were delivered can be used for other projects.

This supports the use of Agile to drive a DevOps organisation structure. The use of the SDN APIs to communicate and to deliver the structure will ensure tighter controls on the development and will produce a checksum in the end-to-end solutions delivery, in which poor security adherence and poor development techniques will be exposed.

The SDN Strategic Architecture also defines real-time service, infrastructure and OSS management and control capability within the architecture. The advancement of these capabilities provides for additional data to be made available, along with the ability to use outputs from the LNMF. These further support the move to a DevOps environment and provide the new approaches to operating and expanding the environment.

Having access to this information permits for the creation of customisable products based upon a flexible, policy-controlled product set. Checkpoint controls within the Agile environment are key, as delivery within this structure can sometimes leapfrog security and quality. The SDN Strategic Architecture approach and integration of the key management principles supports and offer checkpoints through management applications and the open API capability.

This has the advantage of creating an understanding of the expectations of the other architects, designers and developers across the wider organisation, which increases the accuracy of the solution by including multiple perspectives. Members of Agile teams are empowered to be self-organising and self-managing; operating under the single program vision enables them to act and deliver within a structured manner.

This drives a lean thinking approach: Through a clear understanding of the expectations for the data that constructs the features of the deliverable, specific tasks and objectives can be met. As this data is structured through the portal to the customer - from the device to the resource management system to the service control to the billing records - this ensures focus on the service development rather than on component technology interfaces. It also grants synchronisation of delivery, as each group will understand their objective and will focus on what is required for their work between the APIs layers. Such interfacing data structures help create an understanding of the deliverables being defined when agreeing to the sprint deliverable and its timeline.

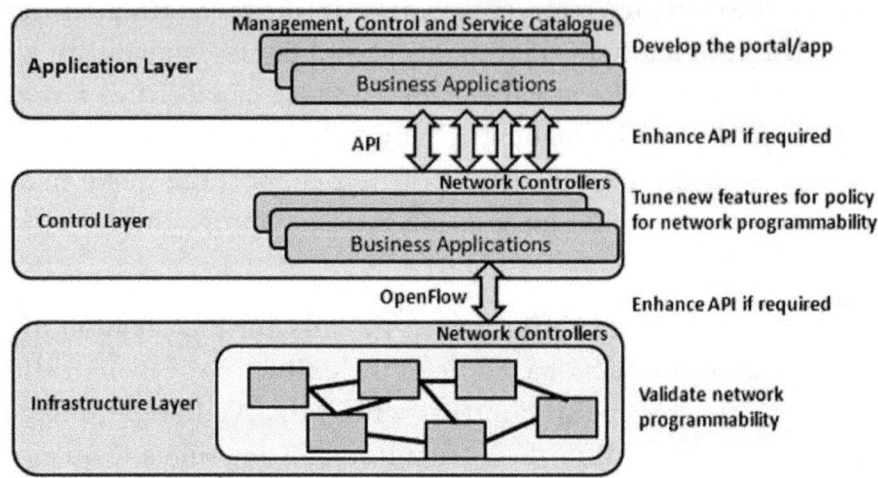

Figure 57: Parallel development using SDN Layering

With the layered approach, no one team can drive changes without the co-operation of the teams whose systems reside at the other end of the API. This ensures co-operative development and makes sure that those who are needed in a project and who understand the system best are included in all discussions.

Such an approach could bring the vital incentives of autonomy, mastery and purpose to the professionals involved, and the ability to expand and understand beyond the predefined boundaries of their roles. Due to the nature of reuse of existing classes, objects etc., this will sometimes offer the opportunity for down-time, which gives the chance to address backlog items, quality, and the opportunity to review the approach that is used.

Summary

The aim of this chapter is to highlight the additional benefits that can be drawn from the evolving technologies and the SDN Strategic Architecture with regard to how projects get delivered, the working practises of the people within the groups and the creation of a business-focused architecture. The intention is to share realisations and thoughts, so that these can be considered when moving forward with these new technologies.

As stated earlier, the thinking for both these topics is not complete. These initial realisations and considerations are shared to allow others to take this thinking further.

13. SDN Controllers

The SDN controller is the technology that creates the centralised focus on the network. Through this centralised point, the controller, the orchestration and management applications can understand, program and control the state of the end-to-end networking environment. The SDN controller is an essential component of this new approach that aims to deliver network and service control. Controller solutions have been developed using both Open Source and commercial approaches, with OpenFlow™ being the most well-known protocol supported for the control of the end device. Much of the development for SDN and OpenFlow™ has been led by the network operators and universities, rather than by the technology suppliers to the industry. This has been delivered through an Open Source approach where not just the concepts of how an SDN Controller could operate were shared, but also the actual source code for the controller. This approach was highly unusual for the networking industry and caused the introduction of a significant number of new entrants into the networking industry.

This Open Source approach has allowed for a variety of SDN controllers to be created, tested and validated by many parties. Each group could learn from the positives and negatives of each approach. This open development approach has created a collaborative learning and development environment in which lessons learned by one were known by all. Through access to the source code and lessons learnt, this open approach has now led to many new vendors entering the market as suppliers to the network operator industry. These suppliers now include many household names from the technology development world along with new start-ups. With such an increase in technology providers comes the opportunity for the creation of solutions that can be used to focus on the individual business plans and business goals of specific organisations.

The changes that the network operator- and university-led development have brought have enhanced the networking industry significantly; and as they compete with each other for business, these changes will permit the network operator staff to drive a focus on solving the problems of service control, management and delivery.

Although the SDN controller was first developed to do only the functions that traditional network infrastructure control delivered, new solutions providers have now expanded this functionality to run and control infrastructures, for example for WIFI, DPI, SBC, etc. It is expected that this service focus for SDN controllers will continue to expand.

13.1 Controller Structure

The high-level diagram below provides an overview of the SDN controller architectural structure. As shown above, the SDN controller sits at the heart of this architecture, in the middle layer between the applications and the physical network elements. The SDN controller uses the applications to identify the resources and to provide the orchestration and the abstraction of the instruction set into the end devices.

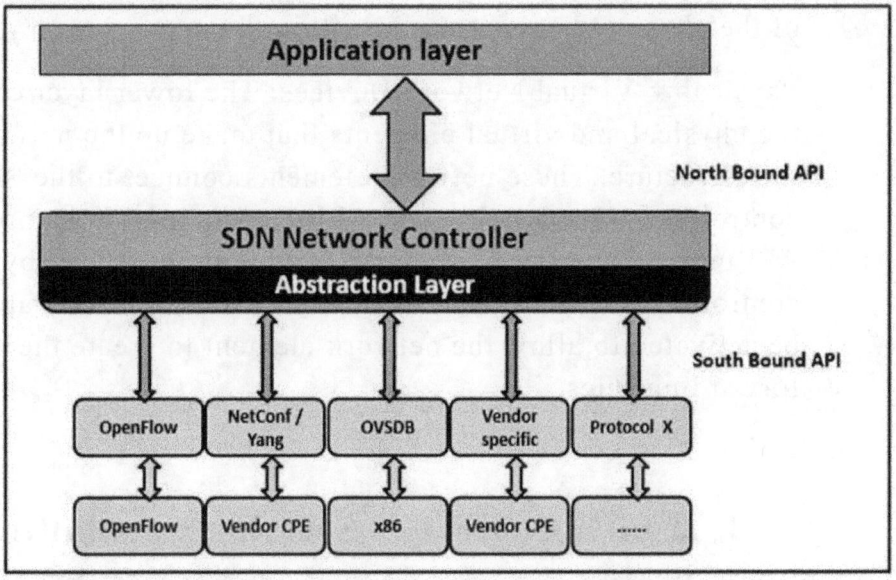

Figure 58: Layers of the SDN Controller

This three-layer model consists of:

1. Network Applications and Orchestration: The upper layer consists of applications that process requests and provide information to identify to the SDN controller what it is required to be signalled into the network infrastructure. The applications introduce business and network logic that is used to create, control and manage a service and network infrastructure according to predetermined policies. In addition, these applications monitor the network and deliver technical orchestration for the service and infrastructure. These applications create a near real time understanding of the state of the network.

2. Controller Platform: The controller sits in the middle layer and provides the framework to create the SDN abstractions. A common set of northbound APIs are provided to interface to the upper application layer. Depending on the protocols supported by the particular SDN controller, abstraction towards the southbound interface will be supported after identification of the capabilities of the end device. One or more protocols are provided as southbound APIs to enable the command and control of the physical hardware used within the network.

3. Physical & Virtual Network Devices: The lower layer consists of the physical and virtual elements that make up the network infrastructure. These network elements connect to the SDN controller via the southbound API. Depending on the protocol or API in use, the network element will be programmed by the SDN controller, or will receive a relevant instruction set that will then be activated to allow the network element to create the forwarding rules.

13.2 SDN Controller Solution Capabilities

At the moment there is no specification that defines the full capabilities of an SDN controller. This has occurred because new opportunities continue to be considered for SDN. Requirements keep growing as operators, vendors and customers realise the benefits this approach can bring to their businesses. This, however, can be viewed as both a positive and a negative. It can be defined as a positive because this way the network operator can now define additional development use cases that they require to deliver their business aims and goals. With controllers being developed in software, rather than hardware, development times are considerably faster than the network operator industry is used to. In addition, this allows operators to experiment and to understand the capabilities of this new technology approach. This allows the business needs to be focused on during the evaluation of this technology rather than just technology compliance that has been the way of the past.

The following, slightly extreme (for most network operators) example is used to explain the variety of choices SDN brings a network operator.

With so many Open Source controllers available, it would be possible for a network operator (assuming the necessary resources are available) to develop their own SDN controller and applications solution, thus avoiding the need of purchasing from vendors. It is recognised that not many companies are capable of this approach - however, this has been raised to highlight the variety of different approaches that can be considered when looking to define a solution that meets the needs of the individual business goals of the operator.

With specification for the network controller lacking, there are no definitive rules upon which the solution could be base-lined against. However, every business has its own business requirements, and these have always been used to generate the technical requirements for any solution installed in the network. Based on these technical requirements, a baseline specification for the solution can be derived.

This book promotes the notion that those professionals who lead the project or the architectural definition are also the ones who control the future openness of the solutions that will enable their business to differentiate itself. To retain openness, this will require that those sourcing the SDN controller define what their business requires, and take a step-by-step approach towards the final architecture.

The following is a list of points that should be addressed when considering which SDN controller to utilise:

- Vendor bundling: Some vendors bundle the solution into a package that includes the management applications, the SDN controller and the network hardware. This approach may work well initially, but should be considered thoroughly in light of future evolution.

- Overlay versus underlay control: Some SDN solutions assume an underlay that is unconstrained in capacity. Identify an SDN solution that drives the strategy and has the protocols and applications that can be used to control the overlay flows on a constrained underlay infrastructure.

- Proactive fault resolution: Identify which use cases are supported for automatic resolution and what is on the development roadmap.

- Controller performance: Investigate the flow setup rates and how the vendor architecture mitigates any of the complex scenarios.

- Modular versus singular protocol controller architecture: Validate the immediate needs of the project and the longer-term needs of the business. Discuss strategic direction and feature timelines with vendors and consider their intentions and longer term goals.

- Modular controller architecture: Identify the protocols that are required for use to configure the current end devices within your network.

- Redundancy: Identify if the SDN controller comes in a redundant and resilient configuration.

- Multi-tenancy: Investigate the ability to control and isolate multi-tenancy environments.

- Network Topology: Identify what the solution has been designed to control and scale to your networks topology size. Many SDN controller solutions are available for LAN, enterprise and Data Centre environments; more are becoming available for VPN and WAN of different sizes and scales.

- Multi-topology: Identify if the controller is able to support more than one topology.

- Multi-service: Identify if the controller is able to support more than one service.

- Identify if more than one controller type is required in your network, what interworking is required and how they will interact/be structured.

- Management applications: Identify all the management applications and functionalities that come with the solution. Identify if these can operate independently or whether they only work as part of a fully integrated solution. Identify missing applications and evaluate the impact this lack of development will have on the automation of the architecture.

- APIs: Identify the APIs supported and the capabilities of the API.

- Identify which vendors' southbound protocols the controller can support

- Hardware interoperability: Identify if COTS and a mixture of legacy hardware can be used for end devices or if the solution requires the selection of specific vendor's technology.

- Identify which analytics are gathered and how these are collated.

- Identify the protocols/APIs used for transport of analytics.

- Identify the migration approach that will be used to drive the achievement of a defined strategy.

- Identify if a few years of vendor lock-in is worthwhile to gain the benefits and advantages of SDN. Better solutions will be available in a few years time, but at that time this will take effort to migrate to and to achieve full feature compatibility.

To understand the technology, it is best to perform proof of concept testing and, from the lessons learnt, create a technology strategy to achieve the business goals.

13.3 SDN Controller: The brain of the network

A term that is often heard around the industry is that the SDN controller is the brain of the network. I haven't been able to identify where this originated from, however it sets the strategy and goal for the evolution of network and service control in a single coined term. With further development, this goal can become real, and the SDN strategic Architecture discusses how with continued evolution, this can become feasible.

The controller with its centralised network-wide view has control of the service and network infrastructure. It can indeed be considered to be the brain of the network, but for the controller to truly achieve this level of sophistication, the controller needs to interoperate with the LNMF and other proactive network management control applications. These applications enable the automation of network and service management, control and delivery, and therefore can over time evolve to deliver an automated solution that receives data, performs the analysis, models the options and identifies a solution (already predefined by network engineers) that can then be programmed into the network.

With SDN, the technology development model that most of us have grown up with changes. The integrated forwarding and control technology approach has constrained both the network operators and the technology vendors. Now those in the lead within the network operator are presented with the opportunity to steer the technology direction, and to create a solution that achieves the goals of their company's business plans using either internal development/Open Source or by using partnerships with vendors.

To attain a brain-like capability, analytics must be gathered about the expectations of the service and on how the service is operating. The SDN architecture achieves this through utilising APIs. The APIs receive generalised information sets from control and business logic systems that are then abstracted to relevant southbound APIs/protocols, for the configuring of the underlying infrastructure. With SDN controllers maturing, applications are increasing, and these now include capabilities that perform a variety of network management, control and support functions.

Some SDN controller network management and control system applications already include capabilities such as device inventory, resource inventory, resource control, network modeller, inventory of the capabilities of devices in network, policy controller, collection of network statistics, etc. The difference to historic network management and control and support systems is that they collect and analyse aspects of the network and service in real-time.

By making these capabilities available, this now allows for additional proactive network management capabilities to be added to the controller. This changes the role of the controller from being a system with a programmable abstraction engine function to being a solution that allows the controller to become the "brain" of the network. These applications are key to the end-to-end architecture solution, and the inclusion of various functionalities enables the delivery of the differing requirements to different areas of a network.

Analytics gathered from the network for service control ensure accurate updates to the infrastructure and utilisation of algorithms to drive proactive network and service management. These capabilities transform network management: What used to be a reactive and offline management solution can now become an automatic and proactive management solution. This fundamentally changes the way of thinking about an end-to-end network architecture.

13.4 Evolving Discussions on SDN Controllers

The applications and protocols supported by controllers heavily influence the end-to-end network architecture and the migration strategy to achieve an SDN Strategic Architecture. Within SDN, OpenFlow™ looks to offer direct programmability of equipment, whereas other protocols look to leverage existing infrastructure and to use protocols such as NetConf/YANG. Although both move towards a similar direction, they don't achieve the same results; with this, it becomes very important that those in the lead within the network operator take ownership of the requirements to create a strategic solution that meets the needs of the business they work for.

However, those in the lead also must deal with existing sunken investment made by network operators – after all, very few companies can consider replacement of this investment. Therefore, the NetConf/YANG protocol offers a migration strategy that permits for the utilisation of existing hardware and allows the realisation of some of the benefits. Again, it will be for those in the lead to steer the development of such a protocol to achieve the same goals as an OpenFlow™-controlled architecture. Additionally, another approach is I2RS, which is a halfway house between fully centralised and distributed. This protocol looks to continue using the traditional routing protocols to perform distributed routing, and influence forwarding decisions with the use of applications. These approaches come with their positives and negatives, and they require extra features/protocols to deliver the data for analytics as they don't yet support this capability natively.

It is the role of the architects within these organisations to safeguard the openness of the industry and to make sure that a successful migration is feasible. This could be achieved through their selection of migration technology sources. For this to be successful, it requires the architects to work with both the Open Source communities and the vendors to achieve the desired solution.

The implication of this is that strategic alliances with Open Source communities and vendors will become significantly more important than they have been in the past.

In the controllers' purist form, the OpenFlow™ protocol separates the network control plane from the network data plane, thus enabling the creation of a single network control policy. This can then be used to program the FIB for the relevant network devices in the infrastructure.

Through programmability, OpenFlow™ enables the controller to remotely instruct the provisioning of the network device. This is done through setting the forwarding tables, using a common instruction set. As a result, operators are able to provision services into the network without having to rely on having to deal with proprietary CLI. This relies on no longer hard-coding a service into the physical infrastructure, as the opportunity now exists to use an overlay model. Historically, services have been hard-coded into the network using complex protocols, such as MPLS L2, Point-to-Point, Layer 2 Multi-point or Layer 3. Although these technologies have gotten the industry to where it is, they now retain significant complexity for service management and creation and constrain the expansion of services. Operators are forced to stay on-net unless limited business partnerships are put in place, and, where management and control visibility is constrained, using NNI agreements.

The change from infrastructure hard-coding to overlay with controllers significantly changes the approach to service delivery. The methodology of hard-coding a service into the infrastructure implies that the service itself is now defined to be those lines of configuration. The per-device configuration of network equipment has been organised on the device in a per-protocol and hierarchical basis to ensure minimise and structure configuration. This allowed configuration blocks to be reusable where possible, which has led to the dependency that, if one of those services were removed or a minor modification was made, other services could be accidentally impacted. This impact can still exist with NetConf/YANG or I2RS. However, a fully developed OpenFlow™ solution could provide the ability to enforce changes with knowledge of the consequences: through a centralised controller, and through utilising application resource and service management controls. The knowledge of the consequences and the use of a layered architecture permits for a management structure to be put in place that allows, for evaluation of the implications and for tuning the changes to meet the needs of the granular service control.

13.5 SDN Controllers Overview

The following information is supplied with the caveat that it is impossible to complete an overview of all SDN controllers that are available today or to record accurately and precisely the current state of the development. This is because with so much development taking place, and with SDN delivering control in software rather than ASIC, a broad array of new features are delivered very quickly from a multitude sources. In addition, so much sharing has taken place within the Open Source model used to develop the thinking around SDN – that it is sometimes impossible to identify all of the originators. This makes it nigh on impossible to ever have a fully up-to-date record, so therefore this information is included to highlight to the reader a widespread high-level overview. Heart-felt apologises are offered by the author to all those who are overlooked, for incorrect allocation of origination of development or for any perceived or actual inaccuracies in the details below!

This overview is given in an attempt to explain that many individuals are involved, who are doing so much excellent work and who are driving change to help the industry evolve to better suit its goals.

Open Source SDN Controller Overview

The following list is not definitive nor does it list all functionalities for all vendors or Open Source controllers. With software development being so much faster than hardware development, vendors quickly add new features and functionalities, and therefore any attempt to create an accurate list is soon outdated. The following information has been constructed only to aid the reader to understand the types and scale of the functionalities that have been and are being included into controller environment, and to help identify the source of many of the concept controllers that are changing the industry.

NOX

NOX is widely recognised as the first SDN Controller that was developed and has been widely shared. It was developed by Nicira Networks and developed alongside OpenFlow™. Nicira Networks was acquired by VMWare, and NOX was donated as an Open Source controller to the SDN community. Since its initial development, NOX has become the building block for many subsequent SDN Controller solutions. NOX could be considered to be the basis for ONIX and POX controllers. NOX has played a pivotal role in the evolution of SDN and OpenFlow™ and it continues to evolve through work done at the University of California, Berkeley. The older branches are now referred to as NOX Classic, and these are known by the project releases as Zaku, Destiny. These code releases have been super-seeded by more recent evolutions that are known as the "Verity Branch". The Verity Branch drops some components: Python support, and that it's not backwards compatible to earlier releases of NOX. NOX has become a C++ developed controller that is targeted at developers who aim to create an efficient and fast controller. NOX supports OpenFlow™ as a southbound protocol. NOX is an Open Source controller that has been shared with the purpose of aiding others to develop their own SDN controller.

Beacon

Beacon is an Open Source project that was first released in 2010. Since then, it has been used as a building block for other controllers. Beacon is a Java Based OpenFlow™ controller and is supported across multiple platforms. Beacon has been designed to be dynamic and for ease of application interfacing. Beacon supports OpenFlow™ on the southbound and RESTful on the northbound.

POX

POX could be considered a second generation of the NOX controller from some of the same team at Berkeley and ICSI. It is targeted at research and university experimentation groups and has been written in Python for flexibility and to allow for the use of a component creation framework. It has evolved based upon lessons learned from users and universities during the development of NOX. POX supports OpenFlow™ on the southbound and RESTful on the northbound.

Open MUL

MUL is an Open Source OpenFlow™ controller that is designed for performance and reliability for live network environments. It is a C based controller which is multi-threaded. MUL supports a northbound interface that is used for connecting to applications.

ON.LAB FlowVisor

FlowVisor was developed by Stanford University and has been widely used by research and education groups. It was created to support network slicing, as was the original goals of SDN. This goal aimed to permit those looking to experiment and to evaluate network scaling and new protocols to be allocated a slice of the network upon which to test and evaluate the theories and concepts upon which they were working on, with minimal impact others who were utilising the same infrastructure. FlowVisor gives the individual and/or groups who are working on a project the ability to manage and control a slice of the infrastructure using their own Network OS without impacting others. This controller was put live on the Stanford network.

ON.LAB Open Network Operating System (ONOS)

The ONOS project aims to build an Open Source distributed Network Operating System. It is being designed to be suitable to scale with performance and availability and to be suitable for service provider networks. ON.Lab has demonstrated an early prototype of ONOS in 2013 and work is continuing on this project.

RYU

Ryu is an Open Source multi-protocol SDN controller framework that provides software components with well-defined APIs to enable developers to flexibly create new network control and management applications. Ryu supports protocols such as OpenFlow™, NetConf, etc.. Ryu is licensed under the Apache 2.0 license and is fully written in Python and is an Open Source controller that was developed by NTT Data.

Open Contrail (Juniper)

OpenContrail is an Apache-licensed open source solution for Cloud network service automation and provides the software basis for the commercially supported Juniper Networks Contrail SDN/NFV solutions. It published features include northbound REST APIs, support for standards-based protocols (IETF draft co-authors included ATT, Verizon) and an analytics engine for the processing of operational analytics that are collected from the controller. OpenContrail supports Linux host-based distributed routing/switching and enables federation within and across scale-out cloud environments using a unified L3 control plane (BGP). It enables control and management using system-level data models and declarative policies which program the low-level configuration to simplify and automate Cloud network service delivery. It supports many features required for virtual private/hybrid Cloud environments including NAT, IP address management, policy-based access control, Quality of Service, group-based policy, etc. It interoperates natively with existing network platforms that support proven BGP control planes for L3/L2 VPNs (IP-VPN, E-VPN) and supports emerging protocols (OVSDB), as well as various encapsulation types (MPLSoGRE, MPLSoUDP, VXLAN). OpenContrail supports various hypervisors (KVM, Xen, ESXi) as well as bare metal and container-based scenarios. It is reported to integrate into any open Cloud orchestration platform and is defined to be redundant, virtualised and highly resilient. Plugins for OpenStack™, CloudStack and Open Daylight are already available as upstream capabilities for the respective projects.

Trema

Trema is an Open Source development from NTT. It is an OpenFlow™ controller framework that has been developed in Ruby and C. Its purpose is to aid others in the development of their own OpenFlow™ controllers by making it easy for other parties to get on board with SDN. With this in mind, Trema is distributed with the full source code that is needed to develop an OpenFlow™ controller. This includes basic libraries and functional modules that operate as an interface to an OpenFlow™ switch. Included in the distribution are some applications and a framework that emulates an OpenFlow™ network with hosts. This framework is provided for the testing of any OpenFlow™ controller that may be built by another party.

Floodlight

Floodlight is an Open Source OpenFlow™ Controller which has been created by a community of developers and supported by Big Switch Networks. It is created in Java and is Apache-licensed. Floodlight is Data Centre-focused, supports both physical and virtual switches and integrates into OpenStack™ Cloud orchestration.

Open Daylight

Open Daylight is an Open Source Java-based controller that supports multiple protocols including OpenFlow™. It started as a network vendor-led open initiative that is now run by the Linux Foundation. The goal of the Open Daylight Project is to deliver Open Source SDN functionality in code and blueprints and to incorporate NFV capabilities into the architecture. The first release, known as Hydrogen, was released in early 2014, with a second release, Lithium, in planning. This initiative brings together almost all vendors, many who have allocated considerable resources, understanding and skills to its development. This joint initiative delivers an Open Source platform to build from, and ensures the integration of functionality for the existing vendor solutions. Open Daylight, through its multi-protocol support, offers a migration path to the goal of achieving an SDN-controlled network. For network operators, Open Daylight allows for the examination of the SDN possibilities. Open Daylight software is a combination of components, including a fully pluggable controller, interfaces, protocol plug-ins and applications. The northbound (programmatic API) and southbound (implementation API) interfaces are clearly defined and documented. By supporting open standards such as the OpenFlow™ Networking Standard, OpenDaylight aims to deliver a common Open Source framework and platform for SDN across the industry for developers.

The hydrogen release of OpenDaylight comes with considerable inbuilt functionality and permits for vendors to include additional proprietary southbound interfaces that can be licensed by vendors. These functionalities include support for a northbound REST API and OSGi, and are redundant, virtualised and centralised. On the southbound APIs, OpenDaylight supports OpenFlow™ (1.0, 1.1, 1.2, 1.3), OpenFlow™ OVSDB, BGP LS, PCEP, LISP and SNMP. Additional functions supported include Open Cloud Orchestration for OpenStack™ through the Neutron interface, including support for virtual tenant networks and supported through a management GUI and CLI.

The Controller Platform includes function for Shortest Path First, Topology Manager, Statistics Manager, Switch Manager and Host Tracker, along with Service Managers that support VTN and LISP. The aim of these capabilities provided through the Affinity Service is to allow the creation of a description that identifies the implementation needs and a view on the network loads. This information is then used to enable the provisioning of the service for the consumer. With so many southbound protocols supported, the Open Daylight controller supports a Service abstraction Layer (SAL) to allow for the identification of how to fulfil the higher layer requests, irrespective of the lower layer protocols.

Commercial SDN Controllers

In comparison to ASIC development, software for SDN controllers is considerably faster; therefore per-product feature lists would quickly become outdated. To avoid outdating vendor solutions in this book, no commercial vendor products are named. Instead, SDN Controller architectural solutions are categorised along with a list of the capabilities for similar vendor architectural solutions. It can be assumed that once one vendor produces an application or SDN product architecture, others will soon follow.

The purpose is to share an overview of the variety of solutions and the different approaches that commercial SDN controllers are now taking.

SDN WIFI LAN Controllers for enterprise networks

- BYOD administration of wireless end devices, wireless security control, application-aware Quality of Service, Intrusion Detection System functionality and content filtering

- Virtualised controller, supported on multiple Cloud orchestration platforms, failover, high availability, programmable via service provisioning and administration portals, centralised, redundant

- Analytics gathering and directional steering of flows based upon awareness of quality of connectivity to end device

- Support for northbound APIs

- Support for southbound APIs/Protocols

SDN Data centre controllers

- Common policy and management framework across physical, virtual, and Cloud infrastructures, designed for optimisation of the application lifecycle

- Inventory, application, tenant, topology monitoring

- Northbound APIs

- Integrated Cloud orchestration with application-level control for the network, security, and network services automation with multi-tenant security, Quality of Service (QoS), NAT, IP address management, policy-based access control

- Clustered with active-active deployment support for high-availability and scalability

- Network programmability for network features including Quality of Service, SLA Management, support for commodity hardware, cluster management, integrated Cloud and switch control

- Algorithms to compute current network state

- Optimised topologies both on optical and IP/Ethernet

- Infrastructure resource awareness

- Control and management of the Data Centre using data models to program the command line

- Plugins to Cloud Orchestration systems

- Overlay support through VXLAN, MPLSoGRE, MPLSoUDP

- Data analytics for proactive infrastructure notifications to the operator based on trends and usage analysis

- Interoperable with existing MPLS technologies using MPLSoGRE

SDN VPN controller

- Common policy and management framework across physical, virtual, and Cloud infrastructures, designed for optimisation of the application lifecycle

- Inventory, application, tenant, topology monitoring, a full per-tenant view of the network and service topologies

- Integrated applications including topology map, resource control, service control, resource management and full network management functionality including initial proactive fault resolution capabilities

- Northbound RESTful APIs for connectivity to applications and for configuration from policy-controlled business logic

- Integrated Cloud orchestration, application-level control for the network, security, and network services automation with multi-tenant security, Quality of Service, NAT, IP address management, policy-based access control

- Cluster with active-active deployment support for high-availability and scalability

- OpenFlow™ as a control protocol with OVSDB as the configuration instruction set for the end device

- Cluster with active-active deployment support for high-availability and scalability

- Support for COTS end customer premises devices, including servers

- Support for virtualisation of customer premises product features into Cloud

- Interoperable with existing MPLS technologies using MPLSoGRE

- Interface to customer portal for instantaneous activation of customer changes through policy based configuration control

- Overlay support through VXLAN

- Support for configuration of traditional end devices through SDN modular controller architecture along with a multitude of protocols

- Secure communication for control signalling through the use of HTTPS and TLS (Transport Layer Security)

- Secure boot strapping of end device for automated configuration based upon interfacing to the subscriber database

- SDN Enterprise network controllers

- Control of enterprise environments including WAN, campus and access link and LAN

- RESTful and other northbound open APIs

- Support for COTS and appliance hardware

- Supports modular controller approach through OpenFlow™, OpenFlow™ OVSDB, BGP-LS, PCEP, OnePK, NetConf YANG, Puppet

- Supports automated deployment and compliance checking of network policies

- Applications for the control and delivery of multi-tenant security, Quality of Service (QoS), WAN optimisation, high availability, integrated analytics, policy, and network abstraction

- GUI-based programming for automated and centralised application-level control of the network for security and network services

- Applications deliver traffic visibility for troubleshooting, analysis, reporting, and archiving using a centralised, policy-based approach

- Topology-independent forwarding, network slicing and virtual patching

- Solutions come redundant, virtualised and centralised

- Secure communication for control signalling through the use of HTTPS and TLS (Transport Layer Security)

SDN WAN controller

- Integrated modelling application (LNMF) with offline and online data collection

- Integrated capacity planning, failure analysis, traffic engineering, and network health and traffic trends analysis for computation of network resources

- Application delivers traffic visibility for troubleshooting, analysis, reporting, policy-focused approach

- Redundant, virtualised and centralised solutions

- Secure communication for control signalling through the use of HTTPS and TLS (Transport Layer Security)

- Supports modular controller approach through OpenFlow™, OpenFlow™ OVSDB, BGP-LS, PCEP, OnePK, NetConf YANG, Puppet

- Supports automated deployment and compliance checking of network policies

- Applications for the control and delivery of multi-tenant security, Quality of Service (QoS), WAN optimization, high availability, integrated analytics, policy, and network abstraction

- GUI-based programming for automated and centralised application-level control of the network for security and network services

SDN Controller for Session Border Controller resources

- Manages centrally a group of virtualised or component-based SBCs

- Load-balances traffic on a session basis for resource efficiency

- Utilises the entire pool of SBC resources in all locations for off-net calls interconnectivity.

- Designed to increase operations' efficiency when growing SBC resources

- Common policy and management framework across physical, virtual, and Cloud infrastructures, designed for optimisation of the application lifecycle

- Inventory, application, tenant, topology monitoring, northbound APIs

- Integrated Cloud orchestration, application-level control for the network, security, and network services automation with multi-tenant security, Quality of Service, NAT, IP address management, policy-based access control

- Cluster with active-active deployment support for high-availability and scalability

Summary

The SDN Strategic Architecture fosters a clear focus on service management by permitting a shift from managing the infrastructure as a method of managing the service, to directly managing the IP flows that carry the service.

The SDN controller offers this capability to deliver control and these come is many shapes and guises. These are targeted at specific functionalities that meet the differing aims and goals of various businesses. Additionally, with the changing control needs for services such as video delivery, home security, home network security, critical at-home support for the ageing population , etc., networks now need to be more agile and definite in how management is delivered.

With this in mind, SDN controllers are no longer just for Data Centres. SDN technology and SDN controllers are now evolving to encompass wide area networks (WAN), multi-technology access networks, SDN optical networks, enterprise networks, SDN VPN network services, peering interconnects, voice services and wireless networks etc. With these solutions come applications that support the operations and service management using network analytics gathered through APIs. This data will be processed and used by the controller applications, to generate instructions to achieve real-time management of the infrastructure and service layer. Such an evolution changes management from being largely manual and reactive, to becoming proactive and automated. This development requires considerable development to create the tools that can address OPEX controls, by delivering real-time service management with automatic and intelligent controls.

This approach brings a layered structure to network management, permitting for simplified application integration. In addition, this layer enables the triggering of commands via the network and the policy controller. The applications permit for automated decision-making, based upon real time knowledge and intelligence to enable control functionality, through processing of the data gathered from the northbound API. This offers the opportunity to leverage Big Data for interactive and near real-time management decision-making.

14. Northbound API Overview

In the context of SDN, the northbound API is bi-directional communication channel that enables the communication of generalised instruction sets between the control and management systems and the SDN policy and network controllers. Within the context of the SDN Strategic Architecture, APIs are widely used to communicate data gathered from the network to applications, for the production of analytics and communicating of instruction sets. The goal of using APIs is to enable a new level of capabilities in network management and control and to initiate the move to service management and control.

The API technology structure has been borrowed from the IT teams by SDN. This is a key component in the SDN architecture. Having this capability in the architecture creates a natural layering between the various systems and brings clarity the data structure, the format and the data to be communicated. Such clear structures also help those responsible for the project delivery to identify the workload and the deliverables for team members. This in turn aids the development teams from the different groups to identify the data that is required to be exchanged between systems at each side of the API, with the aim of enabling a flow though solution. In this, the API not only provides the capability to transmit the data between systems but also aids in the planning and structuring of the product development.

This natural layering and separation between systems allows for the identification of clear separation of functions during the architectural and development cycles. This defined structure aids problem identification during the operational running of the solution. It also drives the awareness of the need to have clearly defined functional applications and systems, and why control should be enforced on feature and functionality creep for functional components used in solutions design.

14.1 Open Programmable APIs

The API technology concept has been widely used in IT solutions development for many years, so this is a tried and tested technology. Its purpose is to allow those involved in programming the solution to develop a consistent and reliable model for communicating between the applications and the controllers. Through APIs, the management and control applications can obtain complex data reports from the infrastructure about the current state of a network element or service. Being able to use APIs to access and source data from networking equipment permits for a new evolution in management and control application development for service and infrastructure analysis. This is further enhanced because APIs are bi-directional and the same API can also be used to communicate generalised instruction sets to trigger changes to the service or infrastructure using the SDN controller.

In essence, the function of an API is similar to a standards-based protocol. With standards-based protocols, fixed structures with predefined parameters are used to exchange data and values that represent the current state of the devices involved. Protocol stacks perform computations on the network element and the data is then exchanged between network elements to enable new state to be identified, using a defined and established communications channel. This allows all the systems that run protocols to communicate using a common language. The difference between an API and a standards-based protocol is that an API is a communication channel that has a structure and format that can be defined by those designing the system within defined constraints. This is very different to the fixed structures used in standards based protocols. The flexible API structure is used to exchange data in a predetermined structure that meets the needs of their communication required. This permits for the extraction and transmission of data, either to or from a network element or a management or control application. Through this structure complex data can be gathered, or complex instruction sets signalled to support service and infrastructure control and management.

Network elements have traditionally shared management information through protocols such as SNMP, syslog, IPFIX, and others. These constrained protocols are required because there are no open APIs available to extract detailed information from the proprietary network elements. The information gathered has generally been recorded on purpose built/designed OSS/NMS solutions. In addition to the usual logging of alarms and events, these protocols extract and record data in an attempt to discover the state of the service and network. The limitations of these protocols created reduced what data could be extracted, and in turn constrained the ability of the network operator to create OSS and NMS systems that could significantly advance service and infrastructure management. With the inability to obtain sufficient data on the end-to-end service state this, in turn, has limited the network operators' ability to be dynamic in the products and services they can sell and how they support their customers. The old adage of there being no point in building a solution which cannot be managed or billed is particularly relevant when considering APIs within an SDN environment. As they provide the opportunity to expose greater data and can signal complex reconfiguration instruction sets to SDN controllers, this creates considerable positive business implications.

14.2 API Business Enablement

APIs are not standardised but do have a defined structure. Access to new levels of management data collated from the network elements enables network operators to develop new products that promote differentiation in the market through new levels of service management or incorporation of other controls. This way, network operators can use an industry-defined building block to create a communications channel that suits the needs of their business. This has the additional benefit that each group involved in the development of the solutions has a clear understanding of what needs to be delivered by them, and what they will be required to process. This allows for parallel development during the product life cycle and leans towards a more agile approach for service development and delivery.

With the SDN architecture exposing programmatic access into the network infrastructure at centralised point, this allows for the communication of a generalised instruction set via an API to exert business logic control of services and network infrastructure. This gives the application programmer the point of reference to interconnect with a very distributed environment, and makes sure that the instruction set they communicate affects the network infrastructure through the abstraction/programmability capability on the SDN controller.

This structure is considerably more accessible for a programmer to use in comparison to using SNMP or CLI. Through this centralised point, a view can be obtained on the end-to-end IP service, therefore ensuring that programmability of the IP service is no longer limited to a per network element view. As an IP service is delivered using communication across multiple network elements this requires the management and control of a set of flows that traverse across multiple network elements and systems. As APIs can obtain data about the service at different points of the network, instruction sets can be communicated into various network elements to ensure the management of the end-to-end service. An example of the real time affect of APIs is that management applications such as the LNMF can now be interfaced to the SDN controller. This API would permit for the communication of generalised instruction sets which would be abstracted by the SDN controller and pushed over the relevant southbound protocol to the appropriate routers. Such changes could include the instruction to re-route a Segment Routed path to an alternative circuit based upon off-line modelling calculations. This may be because the management system has identified a growing number of errors on the link, a trend towards congestion or one of many other conditions. This data could be identified from data gathered from the network elements and processed by a network management application.

APIs do require an in-house agreement on the structure. Additionally, the security teams need to have a clear understanding on what information is being passed across the API, as poor design can lead to the exposure of restricted data.

This leads to financially beneficial opportunities for individuals and for new start-ups to create the great wealth of APIs that will be required for use between different systems. Individuals, existing companies and start-ups could create APIs and register them on an API store website (much like an App Store for applications and games used by the various suppliers for tablets). This would allow companies who require a specific API to lease or buy the most suitable and relevant APIs from a store. Such a situation would avoid companies from having to develop APIs internally, and this approach would speed up project delivery time. This, in addition, creates wealth for those who take the initiative to create the relevant APIs. Courses are already available with some universities along with considerable online documentation and courses available.

14.3 Application Interfacing (Northbound)

Currently in SDN, the REST API is one of the most commonly used northbound APIs being used. As evolution continues and new requirements are considered, it is expected that the number and types of APIs in use will grow.

REST is a natural fit with the SDN controller solution as it provides for the ability of the communication of specific generalised instruction sets. The northbound controller REST API interface is used for the communication between the controller and the applications of generalised business logic. The REST API utilises the same hypermedia Internet technologies and protocols which have been used in web development.

The REST (Representational State Transfer) API is not a standard, a protocol or a programming language. The REST architectural style was developed by W3C TAG (Technical Architecture Group) in parallel with HTTP 1.1 and is based on the HTTP 1.0 design.

REST focuses on the roles of the components, the relevant data structures and the constraints that are required to be considered when interacting with other components. By focusing on these and not on the details of the component implementation or on its component programming, REST allows for a single generalised instruction set to be communicated, which can later be abstracted to the appropriate component level and protocol syntax by the relevant controller systems.

REST API architectural properties

Constraints have been applied to the REST API architecture to enable its architectural properties and to enable any REST API to maintain its architectural properties. Architecturally, this communication will be supported by a policy controller for the translation of business or application logic to the SDN network controller. The SDN controller then abstracts the generalised command instruction set received via the REST API function for programmability to the control plane of the intended network element.

When considering the needs of northbound API to or from the SDN controller, the properties applied to the structure of a REST API satisfy many of the needs of ensuring the future proofed nature of SDN. These are as follows:

Scalability: REST has been designed to be scalable, and this has already been proven by the success of HTTP on the Internet.

Performance: High performance is a requirement in the SDN/NFV architecture. Situations such as request for IP flow-based service Quality of Service from vCPE environments or programming of Segment Routing flows to move from congestion management to congestion avoidance would require high performance. When considering such future capabilities, the REST API as it has been scaled to deliver high performance with web interactions, ensures that its performance capabilities will meet the scaling needs of SDN.

Simplicity: As the SDN controller (e.g. Open Daylight) is responsible for the abstraction/programmability on the southbound API, this ensures that the northbound can remain generalised in its instruction set.

REST API constraints

The six architectural constraints defined for an API to be RESTful are as follows:

Client-Server: This common approach is used to maintain and isolate the controller from the client (systems/applications), issuing the instruction set. This structure ensures that every system that wishes to communicate with the controller has the ability to do so. This creates a layered approach and permits for a uniform interface to communicate through.

Stateless: All states are required to be kept on the client (application) side. The server (controller) does not maintain data about the client state. This results in a much more efficient SDN controller.

Cacheable: Where relevant and for scalability reasons, the client should cache a local copy of information that is commonly used. This improves performance as it minimises the number of times an application needs to query the network's REST API.

Layered System: The REST interface just like the SDN controller utilises layering, to structure the complexity of communications between the systems it communicates with. The use of a layered structure helps simplify the interconnection and allows functional responsibility to define the goals of the application communication. This way the REST API is used to interface cleanly between the two systems within the architecture.

Uniform Interface: Key to building a REST API is that the data returned is presented in the same way.

Code on Demand: This is the only optional constraint of the REST architecture. It operates by allowing for customisation of the end device functionality of a client by the transfer of executable code.

In addition, a REST API request uses HTTP headers to carry out actions such as filtering, ordering, authentication etc. Each API call must receive authorisation, so therefore must be authenticated. To achieve this, an authorisation header must be sent with each request.

Summary

APIs are crucial to the SDN Strategic Architecture, because they ensure that layering and functional adherence is structured and controlled within the architecture. This building block gives clarity to the solution build for both those who design and operate a service. It provides a clean architectural approach for structuring the functional components and a DevOps model.

15. Southbound Protocols or APIs

The industry has long attempted to solve the problems, complexities and inaccuracies of network element configuration and automated service provisioning. Many methodologies and systems are in place, some of which have promised to deliver the silver bullet to resolve service management.

As SDN separates control from forwarding, the protocol or API can be used to affect the overlay or underlay infrastructure for network management and control. To enable control, knowledge is required about the state of the infrastructure. This requires that data is gathered, and for this, a bi-directional API is needed to help transform the industry.

15.1 Automating configuration and analytics

Configuration and provisioning requirements differ significantly depending on whether it is a core network element or a customer premises device that is being reset. However, for many network operators provisioning is not flexible, and current systems constrain the operators in their ability to quickly change direction and to adapt to changing business needs. This lack of flexibility is caused by many factors, including device software version, device type, train of software, resource identification, proprietary CLI, to name just a few. These functions and variables, along with the many others not listed here, are required to be identified to position the correct configuration onto an end device. This has led to network services today being provisioned using either a manual model or a top-down model.

Within the manual model, the most prevalent design model is designing for the one-off situation when customer requirements seem to be unique. In this situation, requirements are gathered, defined, designs prepared, validated and tested, and then an engineer uses a few techniques to apply the configuration, such as the use of scripts, copy-paste or the direct configuration approach to program the end device.

In the top-down model, an application initiates requests for network resources to network inventory system. The network inventory system is required to be in-line with the resources allocated on the network. The appropriate network resources are identified, and a configuration manager then identifies a pre-defined configuration snippet based on the relevance to proprietary device types and their configuration. Next, the application attempts to orchestrate an end-to-end service solution into the network for the various resources required in the network. Other steps required include the back-end testing of software load per vendor, configuration snippet preparation, testing of the proprietary hardware, maintenance and updates to the various NMS and OSS systems etc., which have for the most been prepared as either proprietary systems or created in isolation for the real world vendors as multi-vendor systems.

These relatively static approaches require significant project investment to add change, to change a process, and to adopt new capabilities.

SDN Approach

The SDN approach utilises some of the benefits of the top-down approach, but comes with significant enhancements. The SDN network controller utilises southbound protocols or APIs to deliver the configuration or the programming of the forwarding table on the network element and to gather and report data on service and infrastructure. In addition, a northbound API policy-controlled approach permits for the communication of the appropriate generalised instruction set that identifies the relevant configuration to be programmed or abstracted to the end device and service. This command set is received by the SDN controller through the northbound API and is abstracted or programmed by the SDN controller, using the relevant protocol/API to the appropriate network element(s).

Network infrastructure resources can be requested and allocated from the resource management application at the application layer of the SDN controller, to permit per-service and per-element configurations to be generated. This delivers the capabilities to unify all vendors' CPE solutions through an open configuration approach, using protocols such as OVSDB, OF-config, or by enabling the configuration of vendor solutions through other protocols such as NetConf. Depending on vendor support, the NetConf/YANG model can be used as a migration model to SDN, thus ensuring the reuse of current network equipment and avoiding loss of investment. This provides a very practical and valuable capability for network operators, and, with the development of OpenDayLight controller, it ensures a transition method to full SDN without having to lose any historical investment.

As with everything, there is more than one mechanism that can be used. At this point in time in the evolution of SDN, the following loosely defined categorisations exist. As with everything related to the SDN Strategic Architecture it will be the network operator architects and designers, and not the vendors, who will decide the success of the following approaches.

Proprietary southbound interface

This approach is based upon communication between the network controller and the network element using, usually, a new vendor proprietary communication technique. Many in the industry consider this to be a vendor lock-in mechanism over the network operator. In defence of the vendors, most vendors also support alternative open protocols/APIs for their equipment. As is the principle of this book - going forward, it is the role of the architects and designers to decide the importance of openness to their companies' future architecture. It is them who need to decide and justify to their management the cost effectiveness and importance of being able to use other parties' equipment, to successfully meet their companies' business aims and goals.

Open Standards configuration protocol approach

With many vendors now involved in the creation of the OpenDayLight project, it will require the architect for the network operator to carefully select the controller and southbound approach that will address their company's needs. Support for these is based on the controller used; there are many protocols that are becoming available. Protocols such as BGP-LS, SNMP, PCEP, I2RS and NetConf allow for configuration of the network equipment. Although programmability/configuration is essential for realising some of the benefits of SDN, the full benefits of SDN will be achieved when the analytics can be easily gathered and processed. After all, achieving service and infrastructure management is the goal of SDN, not just achieving automated configuration. It should be noted that the greater the number of protocols required will increase the volume of work required from the network operator to create the required instruction sets. The chosen direction for many network technology vendors is NetConf. Many are close to delivering the NetConf/YANG tables. Whether this protocol is a time to market approach by vendors, or the long-term offering, will be for the network operator architects and designers to decide.

Open Standards programming & analytics approach

Where OpenFlow™ differs from these other approaches is that it programs the flow. It is also bi-directional and can be used to communicate the management data from the network element towards the management systems for analysis. In addition OpenFlow™ can be used to program network elements using approaches such as OVSDB. Development is also ongoing on OF-Config, with the possibility of including NetConf.

When considering the benefits that SDN can bring, it becomes a race between Network Operators to see who is the earliest to enter the market with the most efficient work processes in terms of competitiveness, customer support and time to market. However, the additional purpose for this book being written is to share the opinions of the author on developments, which need focus, as the

Network Operator builds an infrastructure that is specific to their individual business goals. A long term objective for the Network Operators designers and architects is, the need to be involved with and to support the parties, who drive the creation of the most business friendly Open standard south bound API based protocols, for the programmability (infrastructure element and forwarding table) and the gathering of analytics from the infrastructure. The Open Network Foundation (ONF) is focused on this ask.

External architectural influences on the SB APIs

For many network functions the southbound API can be perceived as the method to automate the application of the instruction set to a switch, CPE or router and allows for the repetitive use of the instruction set.

Functionality changes that have been introduced through other technology evolutions SDN are also enhancing the creation of a next generation network. The changes include - but are not limited to - the increased availability of carrier grade routers (defined as having a very long uptime, multiple years), the simplification of the protocols used in the network, the virtualisation of network functions and the introduction of the LNMF. These capabilities also provide for a change in direction in optical networking evolution. With this evolution comes the ability for the introduction of a dynamic capacity and for the automatic restoration of point-to-point wavelengths that can be intelligently modelled and instantiated into the infrastructure. For dynamic optical, work is now ongoing within the ONF to further this evolution under the name of Transport SDN (T-SDN). With these capabilities comes the ability to create new infrastructural architectures that can be considerably more effective in delivering to the needs of the business.

With such advancements, networks now no longer need to be created in the dual device hierarchical structure of the past. This has the implication that as much as 50% less equipment may be required, as an optical element can route around a failed or failing core router to terminate capacity on another element. This means that routing tables would be significantly reduced due to the decrease in the number of active secondary paths existing in the network. In conjunction with the introduction of real-time modelling (LNMF), this creates the ability to use application generated advice on how to program flow-based forwarding so that these flows can be programmed into the network through the southbound API. This drives forward the ability to achieve a next generation network.

OpenFlow™

The goals of OpenFlow™ protocol are too separate the network control plane from the network data plane, and to enable programmability of the network fabric. OpenFlow™ uses a central controller to provision the forwarding table of the network device directly. Its open approach gives the architect many new architectural options to as to how they approach the build of the infrastructure they are responsible for. With its central view of the network and its per-flow focus, the OpenFlow™ SDN controller offers new opportunities for service control and management. As new management and control applications become available, this will allow for the application of business logic that could enable the creation of a new level of service management through automation.

Although SDN is now evolving to incorporating many protocols, the first and currently most widely adopted protocol is OpenFlow™. The OpenFlow™ concept is said to have originated in 2008 at Stanford University, version 1.0 of the specification was released in December 2009. The ONF (Open Networking Foundation) controls the development of OpenFlow™. Additional versions of OpenFlow™ with expanded features and functionalities have since been released and this will continue to expand with time.

Through a controller, the OpenFlow™ protocol manages the specification which covers the components and basic functions of an OpenFlow™ enabled switch. This announcement of this protocol enabled the focus of a lot of organisations and companies to reconsider how networking was implemented. As new features get developed, its capabilities increase and its focus is shifting out of the Data Centre.

OpenFlow™ enables an SDN Controller to program the forwarding plane, thus permitting the ability for it to make adjustments to the forwarding on the network. The controller has a higher level of visibility, due to the analytics it can gather and have the SDN controller applications analyse, which permits an alternative method of adapting the forwarding of traffic to the changing business needs of the organisation. Through OpenFlow™, entries can be added and removed on the internal flow table of network forwarding equipment, thus enabling the network to be more responsive to real-time traffic requirements. OpenFlow™ is a software-defined networking (SDN) standard.

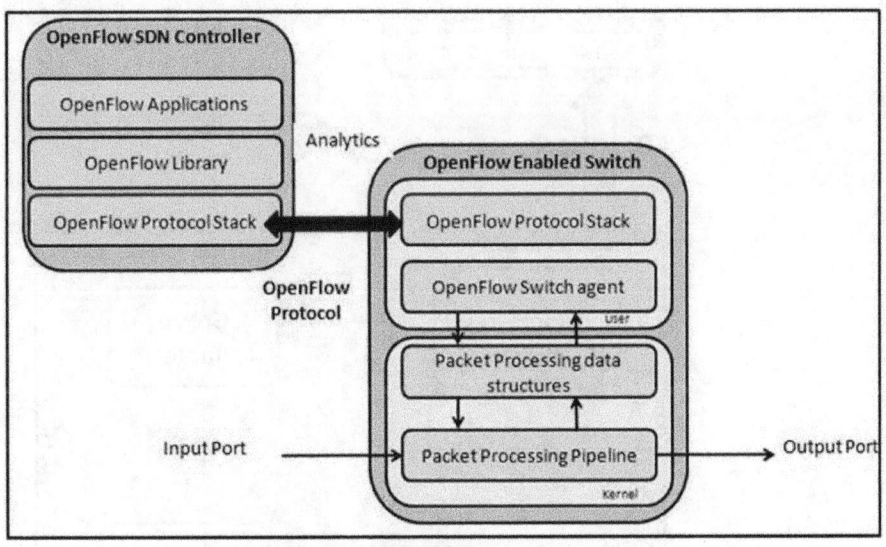

Figure 59: OpenFlow Structure

OpenFlow™ can be considered as a purist SDN protocol. It can receive instructions that have been sent via the northbound REST API, using a generalised instruction set, to the controller. This is then used to create the programming of the FIB. Where the OpenFlow™ protocol is supported on the end device, it directly programs the FIB of the end device. This allows for a very flexible manipulation of forwarding rules to the FIB of the network element. OpenFlow™ overcomes the need for CLI programming and allows for management of multiple vendors' products that support OpenFlow™ from a centralised controller. This allows for the network to be tuned according to the needs of the organisation via the manipulation of packet forwarding by setting rules, which enable adding, modifying and removing packets.

OpenFlow™ Structure

The OpenFlow™ protocol is enabled on both sides of the interface between network infrastructure devices and the SDN controller. As with the services that network consumers use, OpenFlow™ uses flows to identify network traffic based on pre-defined match rules.

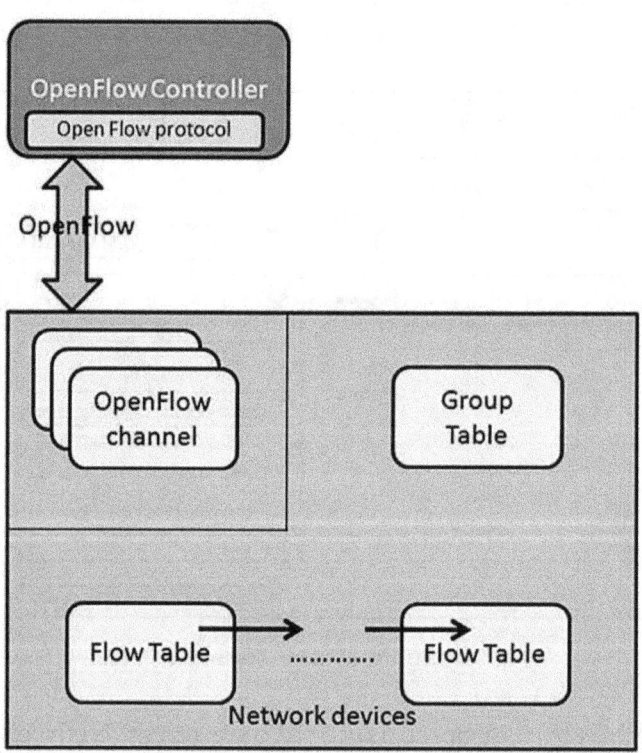

Figure 60: SDN OpenFlow Logical Structure

As features and functionalities get added to the protocol, this will permit for control of the user experience through programming from the SDN controller, based on the user flow requirements, and depending on the service and product purchased and its SLA.

This allows for the management applications to support the control of the network based upon the end to end view they have for services and infrastructure In comparison to bulk forwarding, which is the mechanism that is in use today in networking, the OpenFlow™-based SDN architecture provides granular control that enables the network to control the network according to the needs of the network operator.

The OpenFlow™ protocol can be used by the controller to add, delete and update flow entries in flow tables. This can be done in proactively in real-time. The flow table in the switch contains a set of flow entries. Each flow entry consists of match fields, counters, and a set of instructions to apply to matching packets. Signalling is initiated from the controller to the network device using single of multiple OpenFlow™ channels. The flow entries, the flow tables and the group tables are structured for communication using the OpenFlow™ pipeline to ensure consistency of communication of any instruction sets.

Open vSwitch

Open vSwitch was initially released in 2009 and has been developed as an open-source implementation of a distributed virtual multi-layer switch. It supports multiple networking protocols and standards. It is designed to enable automation of network programming while support standards based network management and interfaces and protocols. It is defined by IETF RFC7047.

Open vSwitch can be run both as a soft ethernet switch operating in a hypervisor and as the control stack for switching silicon. It is available in multiple virtualised platforms and switching chipsets. It is widely supported and utilised in cloud orchestration systems, by network equipment vendors and is distributed in many operating systems, and it is also already widely adopted and used in some very large environments. It is easy to program and suited to policy and SDN network controller manipulation through its OVSDB interface. Open vSwitch Database (OVSDB) is the management solution used to manipulate the configuration of Open vSwitches.

With its wide adoption by many vendors, the solution is well-tested and supported. New features and capabilities continue to be integrated into the software. As its name suggests, it is an Open software solution, and this is reflected in the fact that it is one of the solutions to be included in the Open Daylight SDN controller. OVSDB, at time of writing of this book, supports the following functionality. This is expected to expand:

- Security: VLAN isolation, access lists, traffic filtering

- Monitoring: Netflow, SFlow, IPFIX, SPAN and RSPAN, Flow_Sample_Collector_Set

- Quality of Service: Traffic queuing and traffic shaping

- Automated Control: Open vSwitch configuration management protocol, OpenFlow™ controller configuration, SSL

- Interface programming: Bridge configuration, Port Configuration, Port Mirroring

- Tunnelling techniques: VXLAN, MPLSOGRE, GRE

Design benefits delivered by Open vSwitch

Open vSwitch has gained wide popularity in the hypervisor environment because it is an open solution, it can be externally programmed and because existing proprietary CLI based network equipment did not have the multi-platform/multi-vendor flexibility provided by this initiative. These same principles are now also giving it significant traction in hypervisors and small network nodes, such as CPE. It can be used in customer premises for VPN and residential internet solutions. It is especially suited for use in both business and home component of a vCPE solution, where the CPE requires reduced functionality and where the Cloud is utilised for applications. This provides the ability to control the traffic between the virtualised machine and the outside world.

With its suitability for use with policy-based programmability, Open vSwitch is very suited to the external centralised control that is needed on hypervisors or business VPN CPE. This fits very well into an SDN Strategic Architecture as this permits many COTS-based end devices - to be used in a design and to be automatically programmed using the same SDN policy and network based controller using protocols such as OpenFlow. The CPE market has historically used proprietary switch/router functions. Using OVSDB delivers the ability to apply changes to the configuration in a very dynamic fashion across many vendors white box hardware. Additionally, it ensures complete accuracy of instruction set and permits for automated proactive management through the analytics gathered.

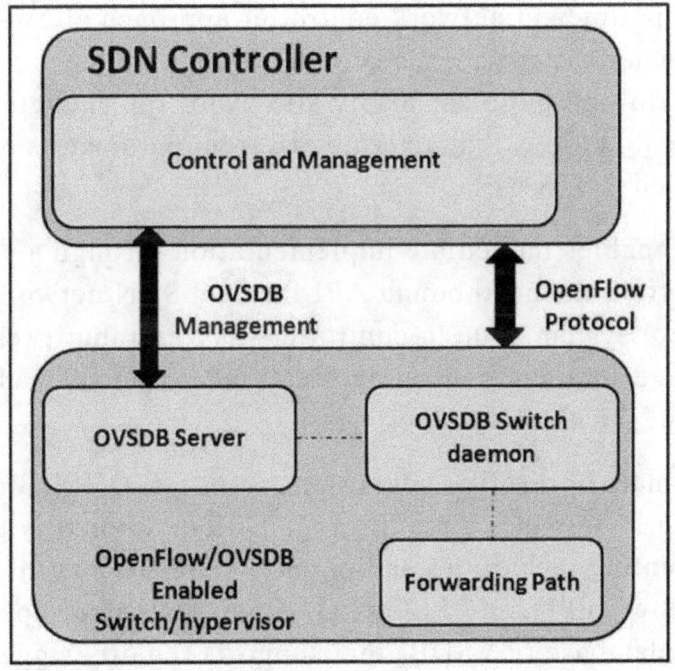

Figure 61: Complementary - OpenFlow and OVSDB

Policy-based configuration instantiation: OVSDB brings a new level of flexibility of control into an environment that has historically been built upon layers of complexity, work processes, long timelines, resources management IT systems and the need for management of configuration across multiple vendors code releases. With the single open command line, OVSDB simplifies the instruction sets required to be identified from a service catalogue. This reduces the need, even in SDN Controller builds for support of

263

proprietary southbound protocols. OVSDB can be programmed using OpenFlow™ as these two solutions complement each other but should not be assumed to require each other. OVSDB can then be used to trigger the configuration onto the device through a policy and SDN network controller approach. With its support for VXLAN and other encapsulation/tunnelling techniques, VPNs can be created without the need to hard-code configuration into transport devices in the network backbone. This permits for the end-to-end use of the overlay approach.

In turn, OVSDB enables customisation using a service catalogue. With its openness and wide support on many vendors' solutions, an SDN policy and network controller approach allows the end consumer to activate the service through automatic programming. This flow-through capability allows for the changing of a fixed set of combinations that activate different parameters on OVSDB enabled COTS CPE.

This enables immediate implementation through a policy control feed from the northbound API from an SDN network and policy control system. Due to controllers having inbuilt resource management and control applications, resources can be allocated as part of the automation process.

Automated proactive network monitoring: OVSDB comes with network flow-based monitoring. This functionality enables simple accounting techniques and permits visibility of flows using Netflow, IPFIX and SFlow. Open vSwitch also supports a network state database (OVSDB) that supports remote triggers - this function triggers actions when issues occur. This capability, especially when fully hooked into the SDN Command and Control solution, allows for a quick response to the issue, for the generation of a proactive management response, and for the generation of a fix to the network problems. This brings a very flexible control of both hypervisor and CPE environments through OpenFlow™, and is especially suited to virtualised environments where instances are being constantly spun up and down.

Multi-vendor hardware support: Open vSwitch's forwarding path is structured to be open to offload packet processing to hardware chipsets. This architectural approach suits both a hardware switch design and a host NIC, and is suitable for dedicated bare metal switches or virtualised environments.

NetConf/YANG

NetConf and YANG are two separate but complementary standard based protocols. In summary: YANG is the data modelling language and NetConf the protocol for the application of the configuration. Although NetConf has been in existence for many years, it has, up until now, not been widely considered for adoption or use by operators.

YANG was developed by the NETMOD working group in the IETF. This data modelling language was first published as RFC 6020 in October 2010. NetConf was developed in the NETCONF working group in the IETF. The NetConf (Network Configuration) Protocol is a network management protocol developed and standardised by the IETF, first published in as RFC 4741 in December 2006, and later revised in June 2011 as RFC 6241.

NETCONF does not remove the native proprietary CLI as it essentially mimics this solution. NetConf/YANG is gaining popularity amongst the technology vendors as it can be used for the programming of instruction sets on top of existing operating systems. For this reason it is used by some operators; it is usually capable of being loaded into existing in-field systems and therefore enables fast movement towards using an SDN strategy.

Whether it will become a long-term protocol of choice for the future is for the architects working in the network operators to decide. NetConf was initially defined to overcome the limitations of using SNMP or programming via the CLI. When compared to SNMP, it does bring some extras benefits for network programming, as NetConf encoded with the standard XML tools.

NetConf brings added security and the ability to control configuration placement and enforcement. It doesn't incorporate a transport protocol and therefore requires the inclusion of additional protocols such as SSH, TLS, etc. Another requirement is that the connectivity to devices it controls and monitors are managed by another protocol. Its current limitations constrain the goals of SDN and limit the architecture in its ability to communicate fully analytics data of what is happening on the device to the northbound applications. In this it doesn't yet fully meet the aims and goals of SDN, as it doesn't fully address network and service management. This protocol, however, does provide a starting-off point for the automation of complex network architectures. NetConf/YANG delivers service control by programming the service into the device, but lacks the integrated ability to deliver a fully flexible path to service management with its limited visibility of the service.

NetConf and YANG do bring benefits, as they may be capable of structuring the constantly evolving command lines when the vendors deliver the full set of YANG data models.

In the SDN Strategic Architecture the Command and Control network management systems needs to be able to push and pull instruction sets through the predefined SDN structure to reflect the state and desired state of the equipment that is controlled for service delivery. This also requires reporting of detailed structuring of the analytics data from the network elements to the network management applications.

NetConf/YANG looks to address the configuration changes needed to run a live network. This is done through the structuring and managing of instruction sets - these are applied to the network to ensure that the network is reconfigured. YANG data modelling achieves this by ensuring the complex configuration of resources are clearly defined, thus ensuring parameters can be set accurately. Through the use of NetConf, these parameters are configured into the network to create installation definitions, and to create change through over-writing of existing programmed command lines and deletion of services.

NetConf provides mechanisms to install, manipulate, and delete the configuration of network devices. These operations are realised via a simple Remote Procedure Call (RPC) layer and by using the XML (Extensible Markup Language)-based data encoding for the structuring of the configuration data and the protocol messages. Additionally, NetConf can return event messages using the NETCONF protocol operations structures as defined within NetConf operations. Much like SNMP these events operate in a client server model and can be filtered and controlled. The events are as yet insufficient to deliver the levels of analytics that will be required by applications to enable an end-to-end service focused architecture.

SSH is supported and can be used to ensure a secure exchange. The NetConf requires multiple protocols to be able to deliver change to the network. This protocol-stacking requires NetConf, YANG and an encryption/transport protocol such as SSH/SSL to deliver these full capabilities. They can be structured into four layers. An overview of these is given below.

- Content layer: This layer defines the configuration and notification data

- Operations layer: This layer defines the base protocol operations for the retrieving and editing of the configuration data

- Message Layer: This layer provides the mechanisms for encoding remote procedure calls (RPCs) and notifications

- Secure Transport Layer: This layer provides a secure and reliable transport of messages between a client and a server

Historically, many protocols have existed that have been used to get and set data. These include protocols such as SNMP, TL1, NetConf, etc. These protocols only ever achieved limited success because of the need for each protocol to create its own data model. This complexity has limited their successfulness and usefulness by the industry. The YANG data modelling language provides the ability to define a solution that could be used to communicate complex data model instruction sets.

When combined with the SDN Strategic Architecture, NetConf offers a strategical first step to moving existing equipment to an SDN-controlled environment. Again, it is crucial to the role of the architect that they steer the change in the industry to ensure that the solutions are enhanced to solve the full problems of the network operator, rather than living with the limitations imposed by existing protocols and systems. With network operators in the lead, it may be possible to drive further enhancements into the YANG and NetConf protocols to resolve these outstanding issues.

With the OpenFlow™ model, the forwarding is programmed to the device. This is not the same as with YANG/NetConf, where a complex instruction set is created off-line, included into a data model and then programmed to the end device. YANG/NetConf is much the same as what occurs today: With considerable complexity they automate the role of the engineer and structure the complexity, so as to replicate what network engineers do today. This is not a new approach, as it only replicates these complex data structures, scripts, and the role of the engineer being the application of the command structure for modifications or changes. OpenFlow bypasses this and drives the programmability of the end node.

As a data modelling language, YANG can be used to model both configuration data and network equipment state data. There is still considerable work to be done to define these data models, and as with any CLI, it will require skilled personnel to deliver these complex configurations. It does bring a focus to a single standardised approach, but architects need to consider whether service management can be achieved as the focus is still on the management and programming of the infrastructure rather than on the flow control of the service.

BGP-LS

BGP is the well-known Internet routing protocol that some vendors are using in hybrid type models of software-defined networking (SDN). BGP Link-state (LS) is a levelled and standardised set of extensions for BGP and utilises Java-based implementations. BGP LS has been identified as a potential SDN protocol that can enable the network programmability.

While SDN proponents have focused on OpenFlow™ as the protocol for decoupling the control plane and data plane of the network, others are considering that having an existing standards-based approach on the southbound protocol and using this to achieve SDN is as important as offering the full operational agility and programmability that a SDN architecture offers. In this model, the controller uses BGP as a control plane protocol and leverages YANG/NetConf as a management plane protocol to interact with physical routers and switches. This can complicate network control and service management significantly.

In this, the BGP LS approach does not achieve the same goals that a southbound API protocol is expected to deliver, because BGP-LS is used in conjunction with NetConf-YANG. This permits the reuse of multiple existing protocols and doesn't take on the full opportunity to drive a new architectural end-to-end approach. The full separation of the forwarding and control plane isn't achieved, and when compared against other solutions available in the SDN toolbox, this solution retains a lot of the shortcomings of the historic network architectural approach.

As with all networks, the business requirements are essential. In some cases, the BGP LS approach may be the only suitable solution available until other technologies are fully developed to meet the business requirements. With considerable investigation and a clear understanding of the final goal of the end-to-end architecture, it could enable a controller-based solution to exist in a multi-vendor environment without requiring large-scale infrastructure upgrades. BGP LS has the potential to serve as a stepping-stone to achieve a more complete architecture in time.

In this case, this solution could be part of an evolution strategy to meet the needs of the business. It will then be the role and the responsibility of the in-house solutions architects and engineers to drive the vendors, internal development or Open Source communities to deliver the functionalities needed to achieve a more complete SDN architecture.

When considering the use of BGP LS as a southbound protocol for program flows across the network, it needs to be understood that

BGP doesn't program flows, but operates and controls at a higher level of state: physical and virtual topologies (L2 and L3), security policies, etc. As the controller operates on multiple levels of abstraction, ranging from control of routing and bridging topologies to flow-based forwarding, the use of BGP LS will add additional layers to the SDN stack. This causes a separation in functionality and adds complexity.

BGP for SDN may offer CAPEX savings by allowing network operators to integrate existing networks into the SDN environment, but it probably needs to be considered in the light of a migration strategy to SDN and not as a final SDN solution. With this, it will need extra requirements to be driven into the solution to drive the evolution of the solution.

Path Computation Element Protocol (PCEP)

The PCE (Path Computation Element) architecture and PCEP (Path Computation Element Protocol) are defined by the IETF in RFCs 4655 and RFC 5440. A Path Computation Element (PCE) is an entity that computes paths complying with supplied policy constraints and migrates only the key closed control function of the control plane to a centralised and open role.

A Path Computation Element is an entity that computes paths on behalf of the nodes in the network. It can be a router, a COTS server, part of the OSS, or a virtualised entity running in a Cloud. When a network node needs a path for an LSP, it makes a request to the PCE using the PCE protocol (PCEP). The PCE has access to topology information for the entire network and uses this in path computations.

Path computation is the key control feature that needs to be centralised and opened up by SDN. For full control of the network, the network operator must control the traffic engineering policy, and that is executed by the path computation function. The PCE architecture delivers access to this crucial function in an open and centralised format, while the existing configurable functions remain on the network elements.

Adding PCEs to a network can be carried out via a gradual migration path in which existing network elements only require a software upgrade in order to communicate with the PCE. Accordingly, a network can comprise a mixture of legacy routers and PCE-upgraded routers, although the full benefits of the PCE architecture will not be achieved until all ingress nodes are upgraded. This possibility to use a gradual migration evolution approach may be of benefit for operator networks as it gives the opportunity to investigate options without having to perform large-scale change or investment. With this, PCE retains significant complexity within the network element.

Summary

As can be seen, from the various protocols or APIs supported on the southbound interface, there are many ways of addressing the automatic configuration/programmability of the end devices. However, for most network operators, automatic configuration is only one of the requirements they need to address as a business. Key to the future evolution is the ability to gather and collate analytics, and the unfaltering drive of a service management-focused approach. The core business of a network operator is to sell services, and not to install technology. For network operators to be successful, architects and engineers need to move forward the evolution of these protocols and APIs, depending on the business requirements over the coming years.

16.   Service Flow Support Protocols

Some new protocols are currently under development. These protocols have the potential to have a very significant impact on how architects and engineers develop, manage, secure and control future services. As the technical evolution continues, these protocols have the potential to create considerable flexibility and control. This will redefine how services on the Internet are controlled.

To aid readability of the rest of this section IP flows are described in this section as the TCP or UDP sessions that a user's application generates. The successful transmission of IP flow traffic across the network creates the Quality of Experience that the user receives. Moving to mechanisms that can identify flows with similar characteristics, and that are able to aggregate, control and manage these flows, enables the network operator to scale the management and control of the individual service - and with that, the users' experience.

This section of the document does not attempt to explain all the functionalities of these protocols, but instead describes how these protocols could be used and evolved to deliver this capability for the network operator. Following the principle of this book, it is for the architects and engineers to discuss and to provide the requirements to push the evolution forwards.

In addition, these protocols could be used to share data with other parties who form part of the end-to-end service delivery and who are involved in delivering the customer experience.

SDN policy and network controllers can be used to utilise the flexibility offered through these protocols to control the aggregation and flow of the service. This is particularly important to the business model of the network operator as the market is shifting from fixed products to customisable products.

Currently there is work ongoing on MPLS SR, and there's a proposal for native IPv6 SR. These have the ability to create a focus on flow control through a backbone network. The other two primary drafts focus on service control awareness, through the addition of bits added after the IP packet header when transported through core network devices: the Network Service Header (NSH) and the IP Meta Data extension drafts. Through writing additional data into the extended header they provide the ability to expose information necessary for control of granular flows and enable the ability to deliver service chaining etc. Consideration should be given to evolving these protocols into the access domains of networks.

Information on these drafts is available on the IETF website.

- https://datatracker.ietf.org/doc/draft-quinn-sfc-nsh/

- https://tools.ietf.org/html/draft-rijsman-sfc-metadata-considerations-00

- http://www.segment-routing.net/home/ietf

16.1 Segment Routing

Segment Routing had already started development before the industry started moving toward SDN. Its initial development was focused on creating a simplified replacement for MPLS TE/LDP within MPLS networks. The purpose of this investigation was to allow for the defining of a service focused path based upon special service needs. With the advent of SDN, its potential was recognised and its design was tuned to meet the extra needs of service control and management within an SDN environment. This approach is continuing in its development. A native IPv6 SR approach is also under proposal from some network operators and vendors.

Segment Routing can be utilised to generate separate transport paths for the aggregation of IP flows that require specific management characteristics control and which override the normal forwarding rules. This permits for the separation of services based on the product purchased and initiated by the customer.

This approach is in line with what network operators sell and how their business operates. This permits for traffic that has similar delivery characteristics and SLAs to be aggregated together and steered over predetermined paths. The identification of these paths can be created using IGPs, manual techniques, or, within the SDN Strategic Architecture context, through using the modelling capability of an LNMF. This can provide a congestion avoidance mechanism that can ensure the delivery of the traffic through the network. This approach provides the capability to evolve away from congestion management techniques that rely on ASICs to manage Quality of Service and, to use congestion avoidance techniques. The congestion avoidance techniques would be based upon using the LNMF modelling capability, to monitor the end-to-end load on the network and to trigger service specific Segment Routed flows using the SDN Controller to steer traffic onto paths that meet the needs of the customer service.

By providing this capability at the aggregated IP flow level, and through incorporating the opportunity that GMPLS and dynamic optical (T-SDN) bring to instantiate new optical capacity under stress conditions, this presents mechanisms to generate new capacity in the network and to program the traffic to flow over the new capacity. This supports the strategy of achieving guaranteed continual service delivery as per the SLA.

The following scenario describes a situation where service specific IP flows could be aggregated onto a specifically steered SR flow for service management. Let's assume, for example, that the business model of the operator is focused on providing residential services using a vCPE environment, and the customer portal has been enriched to allow the customer to select Quality of Service. The customer can register their preference for quality via the customer portal. Later, when a customer triggers a service for which they have requested quality, the vCPE application policy engine would trigger the marking of all traffic and the forwarding of the flows into an appropriate SR (bi-directional) flow. This flow would then be forwarded through the backbone and delivered according to the needs of the customer.

In Segment Routing, a bi-directional path is defined and set up across the network. The initial node uses this to define how the packet should be forwarded through the network segments and triggers this path. A segment can be locally significant to a Segment Routed node, or globally significant within a Segment Routed domain. Flow state is only established at the ingress node to the Segment Routed domain. It allows for the directing of a flow through a defined topological path. This could be enhanced to permit it to be triggered and managed by the LNMF, so that corrective action could be initiated by the LNMF, when nodes or links on the path are identified as trending towards end-to-end stress conditions which would affect the service.

Segment Routing, controlled by the SDN controller and advised by the LNMF, can now permit for alternative paths to be considered and programmed into the network during evolving operational situations. The protocol should be able to be signalled by the SDN controller to permit for the reprogramming of the path. This way, instruction sets can be triggered into the network to tune the selected service Segment Routed path, based on live network knowledge, historical trends and an understanding of the service SLA (jitter, packet loss, delay, etc). This can be achieved by the LNMF using current infrastructure state knowledge and its knowledge of expected trends. This would allow the network operators to take enhanced control of how their networks operate, and it permits them to address the SLA for their customer services. This moves network management from reactive to proactive and from infrastructure management to service management.

Segment Routing and NSH/(IP) Meta Data all complement each other, but they are not integrated at a protocol level. Instead, these protocols provide network and service information and expose this information for consumption by the separated control plane. Through the exposure of service (data about data) information, the need for high-end technologies e.g. DPI systems, is reduced, and a simpler policy control mechanism is enabled to deliver forwarding decisions based on the end-to-end service, rather than the historic per-hop and per-device situation.

When NSH and IP Meta Data protocols are utilised along with Segment Routing, the architects and engineers are provided with the functional flexibility to enable selected control and management of flows of traffic through the network. This permits the SDN controller and policy controller to identify the appropriate Segment Routed path through the network. With the exposure through reporting via the NSH OAM capabilities on latency, packet loss, delay and jitter on a per-flow basis, per-hop thresholds etc. can be measured. When a path is detected as trending towards developing forwarding problems, it can be quickly identified by the fault management systems, and the LNMF can be used to model a suitable alternative path for the Segment Routing to be repositioned onto, to ensure delivery of the correct SLA connectivity.

A use case for Segment Routing can be to minimise the maintenance of large volumes of state awareness, and to create related aggregated service flow management. These capabilities, will be tuned though the SDN controller, therefore this approach can provide strict network performance guarantees and deliver efficient use of network resources, while delivering connectivity across the network against pre-defined SLAs.

Not all this functionality will become available within first or second releases of this technology. Therefore it is the role of the architects to drive the vendors they select to ensure that the full potential of the protocols is maxed out. This can be used to expose the control plane to enable the service control that the operations teams and the customers require.

16.2 Network Service Header

The network service header (NSH) is added to encapsulated packet or frame to create a separate service path to allow the SDN Controllers and NFV nodes to be informed and to affect the service. This structure also provides the mechanism to facilitate the (IP) Meta Data exchange along the service path and to deliver service functional chaining. Service focused functional systems exist at many points in the network and existing architecture does not fit well with virtual elastic services. The Network Service Header provides the ability for the service flow to communicate relevant inserted service data along the path of the service and delivers service chaining between functional systems.

Structurally, the Network Service header is inserted after the IP header and at the start of the IP Packet header - within a fixed number of bits that most forwarding engine equipment can read. This addition to IP provides a mechanism to overcome many of the limitations that exist in IP today. This addresses the sharing of information along the path to support the delivery of the service, as it is the service that is responsible for much of the complexity that exists in solutions development.

The limitations in IP today include the lack of information about the service characteristics, and affects how the service is delivered to the customer within the context of what the customer has paid for. These limitations have required the creation and the operational management of complex and expensive architectures, to overcome the problems faced by network operators. Additionally, this has required the development of costly and complex line rate technologies to investigate the packet in an attempt to identify this information. This has created the need to deploy complex, inflexible architectures that are difficult and expensive to scale, and that have limited the ability of the applications programmer to affect the service and the network operator to manage the service. The additional information performs two functions: providing network context information, and providing service context information. NSH focuses on network context information and the format of the additional header information. The service contact information is described in the next section under the heading of (IP) Meta Data.

The purpose of this protocol development is multi-functional. It also delivers service chaining encapsulation to transport traffic between either virtualised of fixed nodes, to enable the insertion of service relevant data which can then be used to affect the service delivery and to expose service relevant information to NFV instances where SDN policy control can be applied. This permits for additional capabilities in the IP protocol that have never been possible before. This could be achieved through the population of the context fields with service-relevant information, so that this exposed data can be carried along the service path.

The NSH solution is expected to be implemented across virtual and physical hardware and in operating systems. As the information is populated on to the IP packet, the NSH-aware flows are transport-agnostic.

Network Service Header focuses on service chaining and, as the name suggests, this provides the capability to request a defined connectivity path through a fixed number of applications. Having this capability to over-ride the destination-based routing of IP networks is vital to enabling the setting of a fixed forwarding route between virtualised network applications. This allows for the full monitoring, management and control of the service, thus ensuring the full controls are adhered to.

NSH Header Structure

The following flags signal the header information:

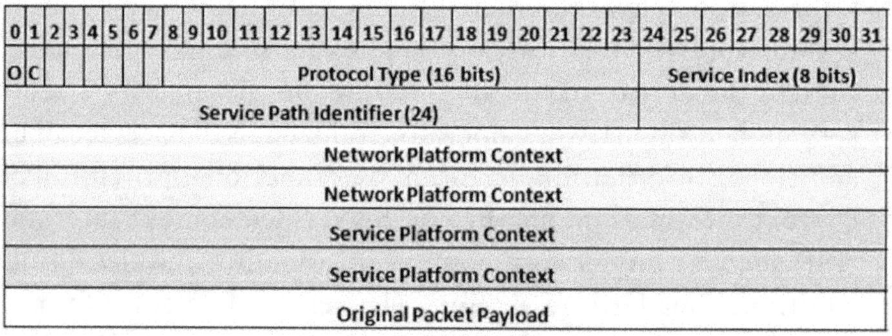

0	1	2	3	4	5	6	7	8	9	10	11	12	13	14	15	16	17	18	19	20	21	22	23	24	25	26	27	28	29	30	31
O	C												Protocol Type (16 bits)													Service Index (8 bits)					
								Service Path Identifier (24)																							
												Network Platform Context																		
												Network Platform Context																		
												Service Platform Context																		
												Service Platform Context																		
												Original Packet Payload																		

Figure 62: Network Service Header (NSH) - Packet Header

- The O bit indicates that the packet is an OAM packet

- The C bit, when set, identifies that the context fields are in use

- There are an additional 6 spare bits which are open for future definition

- The protocol type defines the protocol type of the original payload

- The Service Index can be used with a service path ID to derive unique value for forwarding or provides loop detection and location information, if service chaining is being established

- Service path identifier represents a service path. Service packets are forwarded, based on the defined service path.

Network Platform Context: The Network Platform Context can be used to carry open format information that can be predefined for policy and network control. This field can be used to share information about network devices on the path, or can be utilised by network devices on the path.

Service Platform Context: Service platform Context is used to carry open format information that can be predefined to allow service platforms to exchange information.

NSH Overview

NSH focuses on the providing the capabilities to ensure the control of the flow of traffic between service functions. These are provided through a focus on delivering topological independence for the forwarding of traffic, irrelevant of the defined destination address, to make sure that all network functions are applied to the flow prior to it reaching its destination. This concept is known as service chaining and allows for the service to be created according to the product the operator wishes to serve to the customer. Included in the NSH is the OAM function that provides the capability to monitor and trouble shoot the end-to-end service-specific flow and to report any problems through the OAM messages. When vendors deliver the capability for these OAM messages to be automatically analysed through an LNMF, this will then permit for the ability to reposition the Segment Routed path on the network. This will provide for the network to automatically overcome limitations of the infrastructure path, and thus for automated fault and service-related resolution in an SDN Policy and Network controlled infrastructure. As NSH is linked to the IP flow, it is overlay and agnostic. Meta Data sharing about the service is a key function of NSH and will be described in the next section of this chapter.

Network Service and NFV

Service functions as they are typically deployed in many networks include capabilities such as DPI, security or load balancing. Historically, these have been fixed, and the network infrastructure and the service have had to be designed around the optimum scaling point in the infrastructure where these costly high-end devices had been physically placed. This has created highly fixed networks, and these solutions do not fit with the business model required to enable fast spin-up and tear-down virtualised Cloud services environments. Additionally, this historic model for architecting service functions also doesn't fit well with and an agile DevOps organisational environment operating model, as it doesn't include the flexibility to instantiate services where the capacity is available at speed. This has occurred due to the fixed nature of the equipment, when scaling that must be predetermined, therefore capacity may not be available when it is required and may well be locked down to other currently inactive services. This dedicated deployment methodology leads to a significant waste of resources, which NFV group is addressing through virtualisation of such function.

NSH service chaining supports the flexibility enabled by NFV and Cloud technologies by permitting for the topological independent service chaining. This therefore allows for the full utilisation of fixed or virtual network functional devices.

Using service chaining, the functional equipment can be initiated to support the spin-up of a new application. The IP packets are then tagged to flow through a defined path, thus allowing for the packets to be steered between each functional instance and then transmitted to the end service point.

(IP) Meta Data

(IP) Meta Data is delivered within the capabilities of NSH. It has been created to add contextual information to the IP session on a per-flow basis. This is being introduced, as IP has lacked contextual service information and has historically required the use of extra functional equipment to analyse application traffic.

The introduction of IP Meta Data provides architects and engineers with a new flexible tool set to for the marking and reading of service relevant information. Having this readable data attached to the service flow allows the network operator to apply policy-control to the service without the need for high-end technology systems, and to identify the service on a per-flow basis.

The open context fields that are used to structure this data enables the architects and engineers at the network operator organisation to define locally significant Meta Data for their applications service flows. With SDN separating the control and the forwarding for IP traffic, these combined enhancements allow for policies to be applied to the flows, therefore ensuring its appropriate control and management. Through the IP Meta Data, locally relevant service information is made available to inform the network control as to how the service traffic should be processed and delivered.

For IP Meta Data the structure of the context fields is left open. This permits the originator of the service to include information in the flow that will be needed by the service to allow the control plane to affect how the service will be delivered.

A simple example of this could be to include information to set expectations to the network management system as to how long the session will remain in place, and for how long the network is expected to be busy. This would allow the network Command and Control to plan and schedule the resources on the network in-line with what the overall consumers require, and network monitoring to be proactively aware of the expectations on network resources.

Another example of where locally significant Meta Data would be of value to the network operator would be when a user triggers a movie using a high-end codec. The applications (CDN) transmitting the video could be set to mark the movie with the relevant TV Meta Data. This could include a network operator significant e.g. Hash tag that identifies to the policy/network controller or vCPE environment that the movie is i.e. 120 minutes long and requires a stream that will utilise 12Mb. The edge equipment (the vCPE) will then be made aware that, if the movie is to be successfully delivered, the access network will need to be able to cope with a 12Mb stream for at least the next 120 minutes. Proactive action can now be taken by the access network controller to ensure that the capacity is reserved for the consumer to make sure they receive the stream according to the Quality of Experience that the customer has paid for. This information can be transmitted across the boundaries of networks or removed according to product definitions, or to the decisions of those controlling and managing the networks.

IP has management limitations that create difficulties when managing the session and ensuring it delivers according to the needs of the service the IP flow is carrying. IP Meta Data aims to improve the IP protocol by supplying the relevant information to the intermediate and end devices in the network. This provides data that can enable proactive notifications of what the service requires, and supports the network management systems to control and mange the constrained resources within contention-based networks. This protocol can also be used to communicate service expectations across network boundaries to third parties.

It is expected that this will be supported first in high-end edge network router devices, but its possibilities are greater. It is suitable to be included into video or CDNs, as this will permit the initiator of the session to advise the access network as to the expectations for the services. With an SDN controller, technology developers now have the tools to develop solutions that will permit the operator to signal back and to request a delivered stream, rather than having to utilise ABR (Adaptive Bit-Rate Streaming) or other technologies. This is because ABR has been developed to deliver services and to address the network problems. If, however, a customer has paid for an HD stream, the network should not decide that the customer will receive a lower quality product simply because the network cannot guarantee the proper quality of video delivery.

Additionally, NNI agreements could now be constructed and delivered against for latency etc. sensitive traffic such as SDN VPN, 3rd party video transport, gaming, etc. By using service-specific IP Meta Data that has been constructed using customer Quality of Experience requirements, this permits for service traffic to be steered by the appropriate Segment Routed flow to ensure the SLA is delivered against. With the LNMF having an end-to-end view on the network, it can be used to control the flows based on a few classifications of priority. The Command and Control management system (which includes the LNMF) would need to be used to signal the parameters to be exchanged using Meta Data via an off-line encrypted link. This could be seen as an extension to peering and would permit 3rd party networks to operate agreements to deliver customer requested quality beyond the bounds of the customer ISPs network. IP Meta Data, like most other technologies used across the Internet, will be abused by criminal elements. Rewriting of this information at domain/network boundaries will be necessary, and this can be supported through applications on the SDN controller. Plain language parameter will not be advisable; some form of locally significant hash tags would probably be preferred. These hash tags could be distributed to end devices through an SDN controller application. They could be time-stamped and service relevant. They could also be shared to off-net parties through a secure connection from the controller application, therefore allowing for simplification and security of NNI agreement traffic. They could be changed on a 15-year, 15-week or 15-second level, depending on the decisions by the interested parties as to what is required for the securing of their services.

IP Meta Data provides another toolbox for architects to move away from having to manage the infrastructure in an attempt to manage the service, and to move towards the direct management of the IP flow that carries the service, thus ensuring the service experience for the customer.

17.   SDN Strategic Architecture Transformation

The technologies of Cloud, SDN and NFV used within the architecture have all originated from within the network operators and universities, and were created (mostly by the Web Platform subset of network operator companies) within an open standards-based approach, to address certain business needs of all network operators.

The business driver for these developments was network operator business logic rather than the technology component drivers used in the past. As a natural consequence, these solutions tackle the architectural solution from a network operator business mindset. The SDN Strategic Architecture brings together these technologies and firstly focuses them on the business needs of the network operator, and then identifies a new starting point for developing network operator architectures.

As network infrastructure is the mechanism used to deliver services, SDN with its focus on service and infrastructure management and control, as a matter of course becomes the linking technology between the business systems and the customer, when creating a business focused architecture.

So why is it that the focus of SDN Strategic Architecture has "first" been on the business needs? The technology assembled has been strategically defined with the needs of the network operator and the customer in mind. These new technologies and protocols have been assembled and leveraged to create a customer-focused architecture that delivers according to the demands of the network operators' business model, and not the business model of the technology vendors.

This transformation, driven by the network operators addresses the challenges they face within their business, and this new architectural perspective has already changed the way that network operators are and will continue to structure their architectures going forward.

The author is well aware that the solution defined in this book doesn't suit every business model or demand; the intention is to offer the reader an overview of the authors' thoughts and deliberations on how to achieve an end-to-end business focused architectural solution. In the end it is the calling of the architect or designer to identify and define the most suitable architecture, meeting the needs of their business and justified by their management.

These evolving technologies have learnt and leveraged from developments that have occurred in other areas of the business. They also utilise solid existing technologies and concepts, but has reprioritised these capabilities within the architecture. Many of the technologies that are evolving right now can be assembled and used to create a stronger, more powerful, and, as a result, more successful environment for the network operator – through SDN as the linking technology.

This book reflects the ideas and understandings that the author has been developing with many parties for several years, and that he has been watching and supporting to come to fulfilment. The first step towards making this architecture happen is the acceptance of the need for change. The second step is investing a significant amount of effort to implement the changes, much like the Web Platform companies have already done. For this reason, the SDN Strategic Architecture has been defined to be delivered strategically, with the understanding that upgrades are always occurring within the network operator environment - and therefore change can be introduced gently, without having to rebuild from scratch.

Some of the focal points and goals of this architecture are highlighted in the following sections.

Sourcing of technology for a business focused architecture

SDN, Cloud, NFV, Next Generation OSS and Big Data enable a new methodology for technology sourcing other than the model currently offered by existing vendors. Open Source, internal development and partnerships all become feasible and create mechanisms to deliver the solutions needed. This is possible because of the larger number of technology sources now becoming available that can be integrated through the layered API. This is not to say that vendors are not in the lead for technology development. For most network operators, technology vendors will still be the primary suppliers. This will lead to significant benefits for those vendors who are able to adapt to the changing needs of the operators who is likely to become more solutions focused rather than component focused.

Changing business requirements and customer needs

Customers expect greater service capabilities from the network operator and the Internet. These changing customer demands identify the requirements for the future services and the future revenue for network operators. A more flexible architecture is required, enabling products that address customer needs, and allowing for the many parties on the Internet to interoperate and to deliver the service to the customer.

Web Platform-like approach

Many traditional network operators strive to be more flexible and to emulate the approach of the Web Platform companies. Much of this can be achieved in the evolving technologies, but this will require the network operator to also develop their own products and solutions, and no longer expect everything out of the box. This permits for the creation of individual product creation, and unique management and control solutions.

Layered architecture

APIs provide for the layering of the end-to-end architecture, a structure that allows for the streamlining of the introduction of new systems and for the removal of systems that are no longer delivering business benefit. This approach can work best when systems/applications are designed as functional components with functional boundaries. This controlled functional approach stops the developers from having to include every feature under the sun, which causes issues for the architects and engineers when then need to replace a system that doesn't meet the business needs.

Customer focus

Customer service management and control are at the very core of the evolving technologies. This is crucial to retaining customers, and to lowering the OPEX and CAPEX cost of having a customer.

Unified business focused architecture

SDN, NFV, Next Generation OSS and Big Data provide a new tool set which can interoperate through a unified architecture. The SDN Strategic Architecture highlights many of the key steps required to achieve this. This aims at giving those in the lead of the companies' architecture the tools to ensure its constant evolution.

17.1 Creating new opportunities

To identify the benefits of change, the following two high-level lists have been put together. The majority of the ideas listed can be developed to be provided as real-time programmable capabilities. Please note that these lists just serve as examples and are by no means definitive.

The first list highlights how the technology change can be used to create new network operators services, targeted at both residential and business markets, and how to increase revenue or stickiness through quality. Most customers of network operators are not technologists, so some services listed aim at liberating the customer from having to understand how to make the detailed technology work, and how to ensure it is secure.

The second list highlights possible changes as to how the network operator can run and operate their infrastructure.

List 1: Internet Services for Business and Residential Customers

- In-home managed service delivery

- Virtual ISP networks

- Interworking with the OTT companies

- Managed LAN

- Managed WIFI

- Managed Mobile off-load via WI-FI and fixed infrastructure

- Layer 4 wide area VPNs

- Off-net OTT business or residential VPNs

- Real-time programmable, secure and flexible home working VPNs

- Privacy products that protect the privacy of the consumer and business from overly energetic monitoring

- Customisable web products

- Guaranteed latency, jitter, security to OTT services

- Per-IP flow preferences of a customer prioritised

- End-to-end managed services

- Time-based controls on a customer service

- New search engine models

- BOT-NET tracking

- Accurate and traceable volume-based billing

- Flexible per-flow-based connections excluded from volume based billing

- Selected channels broadcast without quality

- Integrated single service Internet across all access technologies

- Flexible enablement of service-based applications per customer

- Customer applications to be able to be added to vCPE environment

- Parental controls both in-home and out-of-home

- OTT TV channel management and control

- and many others

List 2: Infrastructure and the service mgmt. and control

- Proactive network management

- Automated fault correction

- Greater operational visibility of the live network

- New operational tools to support operations teams

- Inclusion of business logic into security decisions

- Automated root cause analysis on operational issues

- Priority-based per-flow-based management

- An architectural methodology to ensure that those developing new solutions know how new developments should be structured

- Access network control and management

- Unification of multi-access technologies

- Greater content management protection

- Faster time to market of new features

- C&C Quality of Experience interfacing Internet exchanges

- Real-time optical expansion

- P2P optical circuits where capacity is not reserved and is only triggered when required

- Significant reduction in network elements in the network

- Multi-operator backbones with access aggregation, connecting network operators using per-packet accounting, based upon service delivery Quality of Experience requirements

- Authenticated flow-based forwarding to Data Centre, thus minimising impact of bulk forwarding security control systems at entrance to Data Centre

- Exclusion of defined traffic from volume billing

- LNMF multi-layer alternative path selection

- LNMF multi-layer predictive fault analysis

- LNMF multi-layer live capacity analysis

- LNMF multi-layer trend capacity analysis

- Network function resource control, management and monitoring

- Single device carrier grade routers

- Commercial service-focused OSS

- Integrated Big Data architecture

- Reduction of paths needed in the core of networks

- Dynamic optical circuit activation

- And many others

17.2 Enhancing product development

There is little point in creating a great product if it cannot add to a revenue stream or enhance the customer experience. Equally, there is little point in creating a great a product suite if the customer can't quickly and easily understand what it does.

With the SDN capability to deliver policy control instantiation of services using pre-defined building blocks, service enhancement and service development timeframes can be reduced.

With the ability to enable minor changes to restructure a product, the product and marketing teams will now have to be able to deliver faster and more effective marketing communication to the customer. This will need to be done in very down-to-earth language to avoid overloading the customer with unclear combinations and products.

Customisation

Customisation is the option for the customer to select a particular subset of functionality from a product suite. In the process it could be said that, instead of a few products for millions as is the common design process today, the consumers with their own unique combinations create their own individual product, and this, in turn, creates millions of final product bundles for just a few.

It has been identified and is well-recognised that the consumer desires choice, but not an overwhelming degree of choice. Structured choice is possible using the service/product catalogue and policy-controlled enablement of services from the SDN Strategic Architecture.

Faster Time to Market

The vCPE capability is used as an example of a beneficial technology that the network operator could utilise. The following functionality is a list of some benefits that this solution can bring:

- The ability to spin up and to launch new product features within the vCPE framework,

- In many cases, a vCPE architecture avoids the need to change out the CPE at the customer's premises.

- Provides virtualised capabilities and functions can be connected using service chaining thus allowing for flexible product creation.

- Provides the ability to test partial solutions with friendly customers

- Enables the use of Open Source vCPE applications that developers may create now that there is a framework and market to work from

- It gives the ability to gather detailed analytics for proper service management within a Open COTS environment instead of a proprietary CPE

- The use of COTS components, delivers reduced cost

- Policy control for automated instantiation across a large number of devices or for automated customisation of the customer product, and other uses

Friendly Customer Testing

These solution capabilities permit for localised product testing with friendly customers. The vCPE capability provides for the ability to initialise new applications to be spun up on vCPE COTS environments. This methodology of work is very suited to an Agile and DevOps product development and support environment. Feedback can directly be obtained from the friendly customers to guide product development, or to stop the project due to lack of interest.

The SDN Strategic Architectures provides greater management visibility and automatic policy control that fits with the mindset used in a DevOps and an Agile working environment. This fits well with friendly customer testing and general day-to-day working practises.

17.3 Inter-departmental inter-working

SDN breaks the historical isolation of the IT departments with its focus on the integration of the IT Systems into an SDN Strategic Architecture. It concentrates on the Operations Department and Customer Management, providing solutions to enable the effective running of the infrastructure, services and managing customers. It allows the development departments to reutilise historical development and to move to an Agile approach of product delivery by permitting for the clear segmentation of deliverables. This has been made possible because operators have been heavily involved in the creation of many of these technologies, and they have identified and aimed to address the obstacles and issues they experience in their day-to-day working environments.

Vendors of IT systems, Network Management solutions and Network equipment have for years been focussing on expanding the functionality in their systems to resolve all the operator problems and to drive their market share - the language used by the different silos when discussing "their" technologies has evolved in entirely different directions. It is therefore not surprising that, with so many proprietary systems that were isolated, those who dealt with each type of system used a different language that other departments rarely understand. SDN, Cloud, NFV, Big Data and Next generation OSS technologies help to dissolve these artificially built "language gaps", and to change the technology language through the layering of the technologies using APIs, SDN controllers, policy controllers, etc.

As teams need to work closer together to achieve real-time communication via APIs, they will develop greater understanding of the data required by other departments in order to successfully deliver the end-to-end solution. This "build once, and reuse" object-oriented approach enables those doing the development to complete their work, and then move on to do other interesting work.

This will require the sound definition of a clear strategic approach, with an understanding of how the strategy is to be met and delivered. Now is the time for architects and engineers to make sure they not only thoroughly understand their business requirements, but also to keep a close eye on what is happening in the rest of the industry.

A worthwhile exercise to experience and understand how technology can deliver an alternative methodology for service delivery is doing some simple Proof of Concept technology evaluations. This will identify the possible changes. The following high-level list highlights some options with varying degrees of complexity for topics that the Proof of Concept could address.

- Proof of Concept for Virtualised CPE environments

- Reduction of NTU/CPE functionality

- Enabling new application services for the consumer

- Reduction of logistics costs

- Enabling applications in Cloud COTS environments

- Identify improvements in customer experience

SDN VPN

- Investigate policy-based configuration automation

- Evaluate COTS hardware solutions utilising OVSDB

- Evaluate existing CPE solutions utilising NetConf/YANG or other proprietary southbound protocols

- Investigate customer service

- Customisation enabling through automated integration of customer-focused service/product catalogue

Data Centre Proof of Concept

- Integration of Cloud and network management and control
- Identify how automated network functionality can be enabled
- Identify service/product catalogue requirements
- NFV capabilities management and control
- How will automated virtualised network functionality be controlled?
- How will functional systems be programmed?
- How will service chaining be delivered?
- What functions should exist in each functional application?
- Policy-controlled automation of service delivery

Automated network management

- Identify network and service management solutions
- Evaluate policy-controlled solution
- Evaluate APIs and how they will be structured
- Evaluate the analytics returned
- Test and validate LNMF solutions
- How is proactive fault resolution going to be achieved?
- How will infrastructures be managed?
- How will multi-layer control and management be achieved?

Automated service management

- Gather requirements from internal customer service (how they want to manage the customer)

- Gather requirements from product teams (what types of products they want to launch)

- How do the multi-party Internet eco-systems come together to ensure the customer the best experience?

- Identify trend in companies' operational costs

- Investigate policy-controlled automation of service delivery

- Identify network infrastructure and service management solutions

- Identify approach for customer support going forward

- What are the goals of service management?

- What level of security and privacy needs to be enabled?

- Identify what annoys the customer

- Set tasks for the vendors to solve

How and what OSS Systems are required

- Input from the previous two Proofs of Concept help identify what is needed

How can Big Data be used?

- Inputs from the previous three Proof of Concept help build the picture on what is needed

- What architectural and design processes are to be used going forward?

- Which vendors will become strategic partners in the future?

- What does the Procurement department negotiate on with vendors?

- And many others

18.   Synopsis

When considering a move to SDN, there are two possible approaches.

One is: Wait and see, - The other is: Test and validate.

The second approach needs to be looked at with great diligence because the tools created in this technology evolution can be tailored to the individual business needs of the network operator. In the past, all competitors only offered a very similar toolbox to work from, and were subject to similar technology constraints. Today, however, the competitors have a broad array of options at their disposal for creating new methods of servicing the consumer in this highly competitive market.

SDN Strategic Architecture and the accompanying technologies are just as much about revenue growth as they are about cost saving. Looking solely at how further cost can be driven from the network grossly minimises the value that these technologies bring.

Operators who move to these new environments are doing so to gain advantage in a services-based market, strategically looking to leverage value from their network, and to capitalise on this newly created revenue opportunity.

This will be done in multiple different ways, some of which are listed below, in no particular order of significance. This is because every company has different requirements and priorities.

- Operational simplification

- Operational enhancement

- Exposing the flow to ensure the SLA is fully met

- Gathering Big Data to support customer satisfaction

- Enabling product concepts

- Improve time to market

- Differentiation in the market

- Big Data for better customer support

- Big Data for better service identification

- Enabling greater user experience

- Moving to pro-active network management

- Enabling real-time OSS stack

- Enhancing customer experience, hence retention

Moving forward and fostering the development of the SDN (business-focused) Strategic Architecture importantly shifts the focal point to OPEX, CAPEX, customer service, and on the opportunity to create new, compelling services that delight the customers.

These are the strategic and operational goals of all network operators. As such, these are also the reasons as to why much of the technology has been created, and for writing this book.

Continuation of development is expected and, no doubt, required. I do hope, however, that this book provides an interesting overview as to how this technology, which is required by network operators, could be steered and championed by architects and engineers. This is to encourage architects and engineers to take the opportunity to develop a strategic architecture that both meets the network operators' business targets and the expectations of our customers, who simply desire a nothing less than a great experience.

www.ingramcontent.com/pod-product-compliance
Lightning Source LLC
Chambersburg PA
CBHW080635180526
45168CB00008B/3174